Marketing & Selling Professional Services

in Architecture & Construction

Marketing & Selling
Professional Services
in Architecture & Construction

Basil Sawczuk

A John Wiley & Sons, Ltd., Publication

Wiley-Blackwell is an imprint of John Wiley & Sons, formed by the merger of Wiley's global Scientific, Technical and Medical business with Blackwell Publishing.

Registered office
John Wiley & Sons Ltd, The Atrium, Southern Gate, Chichester, West Sussex, PO19 8SQ, United Kingdom

Editorial Offices
9600 Garsington Road, Oxford, OX4 2DQ, United Kingdom
2121 State Avenue, Ames, Iowa 50014-8300, USA

For details of our global editorial offices, for customer services and for information about how to apply for permission to reuse the copyright material in this book please see our website at www.wiley.com/wiley-blackwell.

The right of the author to be identified as the author of this work has been asserted in accordance with the UK Copyright, Designs and Patents Act 1988.

Wiley also publishes its books in a variety of electronic formats. Some content that appears in print may not be available in electronic books.

Designations used by companies to distinguish their products are often claimed as trademarks. All brand names and product names used in this book are trade names, service marks, trademarks or registered trademarks of their respective owners. The publisher is not associated with any product or vendor mentioned in this book. This publication is designed to provide accurate and authoritative information in regard to the subject matter covered. It is sold on the understanding that the publisher is not engaged in rendering professional services. If professional advice or other expert assistance is required, the services of a competent professional should be sought.

Library of Congress Cataloging-in-Publication Data

Sawczuk, Basil.
  Marketing & selling professional services in architecture & construction / by Basil Sawczuk.
    p. cm.
  Includes bibliographical references and index.
  ISBN 978-1-4051-8187-7 (pbk. : alk. paper)
  1. Architectural services marketing. 2. Construction industry–Management. I. Title. II. Title:
Marketing and selling professional services in architecture and construction.
  NA1996.S29 2009
  720.68′8–dc22

                                                                                    2009020257

A catalogue record for this book is available from the British Library.

Set in 10/13 Palatino by Laserwords Private Limited, Chennai, India
Printed in Singapore by Ho Printing Singapore Pte Ltd

1   2010

To Sonia and Luke

Contents

List of Illustrations

List of Tables

About The Author

Basil Sawczuk qualified as an architect. He very soon found the buzz of winning new clients more exciting and rewarding than designing and running projects. After a string of successful wins his colleagues encouraged him to become a full-time job winner, or business developer as it is now called.

At the multi-professional firm of DGI International, he secured projects for many UK blue chip clients and then went on to pursue an international role winning overseas work for some large global clients.

When WS Atkins acquired DGI International, Basil became the midland marketing director and was soon promoted to the national role for the Property Design Division and then as marketing director for WS Atkins Property Services. This incorporated three divisions, selling and marketing architects, civil, structural and building services engineers, quantity surveyors and facilities management operating out of more than 50 UK offices.

Basil Sawczuk has also been a marketing director, and was on the management board of a regional law firm and was marketing, business development and bid director for a major contractor focusing mainly on outsourced public sector work. He is available to undertake consultancy assignments to help your business win more profitable projects. He would be pleased to advise you on any aspect of the seven stage process outlined in this book. Also, he is a speaker at conferences, seminars and training workshops and can be contacted at **basil@potentialise.com**.

Foreword

The ability to effectively sell and communicate is vital to any organisation. Not too many years ago professional firms were prohibited from outwardly promoting their services; now, they compete on an equal footing and their ability to market their capabilities is fundamental to their success. Quite properly, marketing and selling is now high on the professional firms' agenda and many have sophisticated approaches to business winning and development.

Equally, buyers of professional services, particularly the larger client organisations, have developed structured processes for selecting their professional advisors and service providers that they consider best fit for the specific needs of a project and them as an organisation.

However, for many of those involved in professional services, selling is still regarded as a necessarily evil. The reality is that if we believe that we, or our firm, have the right experience and service, we have a responsibility to sell effectively and to be able to communicate with a prospective client to persuade them that we have the right solution.

This book provides a structured approach to identifying and selecting clients, understanding their needs and expectations, aligning to their needs, pitching and winning business. It provides practical steps that are proven and successfully used by its author Basil Sawczuk for over 25 years.

I know from practical experience that the ideas and process that Basil sets out in this book work. They are not rocket science, some points are blindingly obvious, but the combination of the whole and a commitment to each stage of the process provides a good route for any professional firm.

David Jennings
Managing Director of Business Vantage and
Chairman and Founder of the Movers &
Shakers Property Networking Club

Introduction

The construction industry

There are many challenges facing the global construction sector. Most of the developed countries are experiencing similar issues such as the lack of skilled workers, the migration of overseas workers to areas of high demand, increasing costs of materials, increased risks and a greater focus on sustainability.

At the turn of the century, construction activity was booming in most regions of the world with global hot spots being China, the Middle East, India and North America. In 2005, it was estimated that the global construction output was approximately US$4.6 trillion. The UK output was ranked within the top ten in the world with an annual output in 2006 of £113.5 billion. Output would see a setback with the 'credit crunch' in the second half of 2008. The United Kingdom is also recognised for its knowledge economy where design organisations, serving the construction industry, are among the best and largest in the world.

The construction industry is made up of five parts as follows:

❑ The clients which commission the projects. These can vary from large government departments funding schools and health pro-grammes to small businesses having a need to have a one-off investment to improve their property. This could be for new build-ings, or for extension, refurbishment or maintenance of existing buildings. These, and all the various other client types, will have different needs, procurement processes and demands on the other parts of the industry.

❑ The construction enterprises that contract to provide clients with their physical needs. There will be large and small general build-ing contractors, house builders, civil engineering firms and those enterprises focusing on the larger repair and maintenance market.

❑ The construction consultants such as architects, civil, structural and building services engineers, cost consultants and project managers

❑ Other professional services providers which serve the industry such as legal, accounting and IT specialists.

❑ The manufacturers and suppliers of the building materials, components, plant and equipment.

This book concentrates on how the professional services providers can improve their selling skills to the other members of the construction sector.

In a survey carried out by The Construction Industry Council on the UK construction professional services sector in 2005/2006, the following was estimated:

❑ Within the United Kingdom the professional services firms earned £13.9 billion per year;

❑ There were 27 950 professional services firms in the United Kingdom employing approximately 270 000 people;

❑ Of the total earnings engineering firms accounted for £3.9 billion (28%), architecture services £3.3 billion (24%), surveying services £2.3 billion (17%) and management services £1.7 billion (12%).

From these figures, it can be seen that there are many professional services firms chasing a share of work from this large market. In prosperous times, there may be more than enough work to go around, but when times get hard these firms are chasing a diminishing workload. For the successful firms there will be sufficient work in all stages of the economic cycle. Unfortunately, the weaker firms will have to contract by downsizing employee numbers and having to accept reduced profits.

During the slowdown of the construction sector in 2008, driven by the 'credit crunch', there was more pressure on the professional services firms to pursue work from a diminishing market. Those firms that had in place a strategy and wide portfolio of clients were better placed than those serving niche vulnerable sectors. Housing was the main casualty and firms servicing that sector were to suffer the most.

Even in the slow market of 2008 there were large projects being commissioned in the public sector, especially within education. This emphasises the need for a variety of client types within your portfolio. This book will show the professional service provider the process of developing a strategy to secure more profitable work and how to improve success rates.

How this book can help

Many professional services firms serve the construction sector, be it design, engineering, surveying, management or a whole range of other specialist services. The professionals entering these professions spend many years in training getting to grips with the technical aspects of their chosen field. They may also be trained in some aspects of business and contractual matters. Very few will have been trained or even introduced to the art of winning work.

Within these professional services, promotion comes to those who can manage projects and their clients well and also to those who have the ability to bring work into the business.

Bringing in the work or 'selling' was often the territory of the partners or directors. Often the senior partner would be wining and dining current and potential clients. In recent times, there has been the emergence of people whose task is to seek out and win new business. The role of the business development director or manager has been introduced into many medium-to-large firms. In smaller firms it is still left to the partners and directors.

Many professionals find selling their services daunting and do not know where to start. This book will set out a logical approach for those new to the 'sales' arena. The book will also be useful for those already involved in creating new business for their firms.

I have spent over 25 years selling professional services, to the construction sector, within single-discipline practices, multi-professional firms, law firms and for contractors. I have also sold professional services for small-to-medium-sized firms and within one of the biggest in the world. I have discovered that there is a common need for all those firms, no matter what professional service they are selling or what the size of the organisation is. They all required a process, a methodology to make the process more structured. They all wanted an approach that could be followed, no matter how much time they had for 'selling'. Be it a full-time business development director or the associate with a few hours a month assigned to winning work, there is a need to know what works well, how to prioritise your time and resources and how to secure the repeat business.

The need for a process

The construction industry is a complex one. Many people are involved in the creation and implementation of a project. There is the client team that may include in-house technical experts that develop the brief and appoint the implementation team. There are the various specialist professional people who design, cost control and administer the project. Then there is the contractor who brings together all the trades and suppliers to bring the designs to life. On some projects, the contractor may engage the professional team as part of a complete package.

The professional services providers need to sell their services in a way to maximise their resources, improve their success rates and negotiate the best possible financial deal. The client has become more sophisticated, and although price is important, the selection process will involve a checklist that makes sure that the most appropriate professional is appointed. If the professional is equally sophisticated, he will make sure that only the projects that he has a good chance of winning are pursued.

This book will help those professional services firms develop a strategy to secure and maintain an enhanced share of the market. The techniques described in this book will improve their chances of winning work, even when the competition is more experienced and has, on the face of it, better credentials. This is not a short-term fix but must be seen as developing a strategy over a length of time, probably years, to put themselves in a position to secure the more profitable work from clients they want to work for.

Unfortunately, this book will not improve the professional service provider's chances of success if they are not able, or willing, to adapt their services and skills to suit their client's needs. The professional services firm that does not spend time listening to the needs of their clients and tweaking their service delivery to suit, will not obtain the full benefit from this book.

Some professionals believe that they will never greatly improve their chances of success because they are not natural 'rain makers'. They believe that they are destined to struggle and just be satisfied with the peripheral workload that their bigger, or better prepared, competitors choose to ignore. If those same professionals invest their time in developing their services, as outlined in this book, they will be placing themselves in a more advantageous position. If they don't invest the time, they may be reliant on the market to come to them and, as we will discuss in this book, these clients may not be the best ones to pursue.

Those professionals who plan to succeed will have more chance of success. They will create their own luck, they will manoeuvre themselves to maximise the opportunities that may come their way. Of course luck may well play a part in business, being in the right place at the right time. However, good preparation and planning will help to convert opportunities that others may put down to just good luck.

The goal of the professional services provider is not only to win work but also to win that work at a decent fee. It is not too difficult to secure a certain amount of work at a loss, to be the cheapest in town. The aim of this book is to show you how to develop a target list of clients you want to work for, who will value your work and pay you a decent fee for your services. Targeting clients who have a need for services over several projects will give you an opportunity to win repeat work and become more profitable by not having to market yourself from a standing start each time for each client.

By developing these satisfied clients into 'ambassadors' they will be able to sing your praises and refer you to other potential clients, who will want and value what you have to offer.

There is, therefore, a need to have a process to enhance the chances of winning new profitable work and developing the client relationship so as to obtain repeat work.

A seven-stage process

This book will take the reader through a seven-stage process (see Figure 1).

The time span between the start of Stage 1 and delivering added value in Stage 6 can be months or years. A professional firm, with a suitable track record, could submit an expression of interest for a public sector project and within 6 months (or less), be short listed from a pre-qualification stage and go on to win a project through a tender process.

On the other hand, trying to get onto a framework contract for a large corporate client could take many years, especially if they only review the firms on their framework every 3 or 5 years.

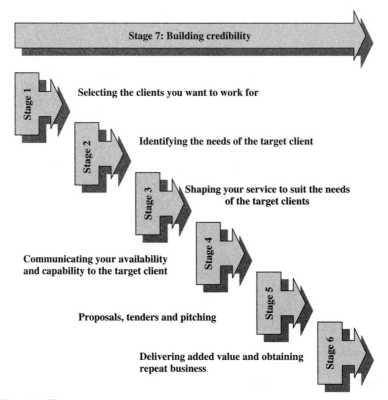

Stage 7: Building credibility

Stage 1 — Selecting the clients you want to work for

Stage 2 — Identifying the needs of the target client

Stage 3 — Shaping your service to suit the needs of the target clients

Communicating your availability and capability to the target client

Stage 4

Proposals, tenders and pitching

Stage 5

Delivering added value and obtaining repeat business

Stage 6

Figure 1 The process.

In other instances, the process could be shortened if, for example, a potential client approaches you and asks for a proposal and appoints you on that basis. The one thing that is certain is that no two projects will be the same. They may have different evaluation criteria and different evaluation teams and have different needs. It is not the same as selling a product. Selling professional services is mainly to do with people. People tend to buy people and their decisions are reinforced by the professional services firm's credentials such as capability, compatibility, credibility and reliability.

Stage 7 takes place throughout the process. It is a continuous effort to build up that track record and credibility to help secure the next project.

Some professional service firms might be tempted to start at Stage 4 without the preparatory work in the previous stages. These firms might well succeed in obtaining the work but by going through all the stages the professional services firm will do the following:

❑ Form a strategy for selling their services.

❑ Increase their chances of success.

❑ Work for more profitable clients.

❑ Improve their chances of repeat work.

❑ Know when to say goodbye to unprofitable clients.

❑ Maximise their valuable time and resource.

❑ Have a better view of their pipeline of potential work.

Overview of the seven stages

Stage 1: Selecting the clients you want to work for

So many professionals are only too happy to work for anyone who may come knocking on their door. In Stage One it is important for the professional to be in the driving seat. The professional should select the clients he wants to work for. There should be selection criteria, proper targeting and courting of the clients. Doing this will help to increase the chances of winning profitable work.

Within the first stage of the process, there is a need to take stock of your current situation. You will need to examine your current and past clients to see who the profitable ones were and who the ones that caused you a loss were. You will also need to see how exposed you are to the market sectors. Are you in a niche area that could disappear overnight or are you well covered across many sectors, both public and private?

It is within this first stage that you will set out a strategy for moving forward. There will be a need to prioritise your limited time and resource to target existing, past and new clients. You will also need to consider your sphere of operations; are you a local, regional or national player? What is your skill set and can that be transferred to other sectors?

You will need to consider the conflicting ratio of effort versus reward. It is not unreasonable to find that the bigger projects with best rewards will be more difficult and time consuming to secure. Also, there will probably be more competitors chasing those bigger projects. Therefore, you will need to balance out your time and effort so there is a good chance of success – perhaps giving some time to pursue the profitable long shots. At least you will be aware of the low probability of success when you set out on that course.

Many books written on selling professional services will not introduce you to the area of lifetime fee opportunities that clients may have. In this first stage, your target selection will be influenced by the potential lifetime value of clients. If you are going to invest time to win work from a new client, then it is best to chase those clients who will have further work in years to come. There is little point in chasing the one project client if more lucrative clients are as easy (or difficult) to win.

Having established your target clients there is a need to build up a rapport with them and find out their future plans so that you can pick the correct moment to pitch for work. No point pitching for work if there are no projects on the horizon for a few years!

All your research is then recorded and a pipeline created, which is regularly monitored and updated. In time, you will get to a stage when you will have a whole string of opportunities with various degrees of success probabilities lined up to convert into projects. The more you feed into the system, the more that eventually drops out as secured projects.

Stage 2: Identifying the needs of the client
If you don't understand the needs of the client, you may waste a lot of time and effort in chasing potential clients who are not going to give you the work for one reason or another. If you know what the client is looking for, you have more chance to provide the bespoke service and increase your chance of securing the commission. Equally, if you find out that you are unable to match the client's requirements, you can move on to the next target without wasting time and money.

Within this stage, you need to identify who the right people are to approach within the client organisation. You will need to know who the decision makers are and who the influencers are. If you don't, you might spend a lot of effort dealing with someone who claims to have influence but doesn't.

Within this stage you start communicating with your target contacts, be it written, telephone or face to face. Each method has its place in the building up of a relationship with the target client.

One of the key skills within this stage is listening skills. You will learn so much by properly listening to your target client. Make a note of what they want, don't want, like or dislike and who is working, and has worked, for them. Your questioning skills need to be fine-tuned to elicit the information you want.

Stage 3: Shaping your service to suit the needs of the target client
Clients need to feel that they are going to be looked after. They want a bespoke service. They want to know that they can work with you, that you have the experience, expertise and resource to satisfy their needs.

There is little point pursuing a client if you are not able, or prepared, to provide the service that they need. Your client will want to think that you are offering exactly what he wants. You will need to shape your offering so that it appears bespoke to the target client's needs. Clients are not usually happy to have to adapt their needs to suit the service that is available.

Take time to reshape your offering so that it is seen as being a much bespoke service. If you are pursuing an industrial project then there is little point featuring your experience in housing. But there would be scope to expand on your knowledge of logistics and how stock is stored and moved if you were chasing a warehouse project.

In all that you do, when describing what you can offer, always see it from the client's point of view. This stage will look at how to tease out the client benefits from any service features you may be delivering. You will need to differentiate and stand out from your competitors.

This stage also revolves around building trust and rapport. When looking at your service offering you will need to establish your capability, credibility, reliability and compatibility. All these are important elements within a trust-building relationship.

Stage 4: Communicating your availability and capability to the target clients

Having satisfied yourself that this is the right client for you and that you can provide the appropriate level of service, you need to communicate this to the client. This may well be a long process, not just months but perhaps years. There will be a need to identify the right decision makers, raise your profile, differentiate your offering form the competitors and develop a relationship.

Through your pipeline development and building rapport with the target clients, you will have established when the next project is likely to occur. You will also have established when the client switches into buying mode. This is where they become receptive to discussing the project and start their evaluation of the potential professional services suppliers. Leave the approach too late and you have missed out. If too early, you run the risk of irritating the potential client or peaking too early. It's all a matter of timing!

Your communications with the target client organisation will be with the decision makers and you will need to have the key influencers on your side. It is time consuming, so you will need to prioritise your effort on those clients who are likely to consider your firm or are in a position to switch, from their current and past service providers, to you.

Having created a pipeline, there is also a need to get around the construction community. There is a need to network; you need to raise your profile. In this stage we address the many ways of getting yourself known to people who can give you work or introduce you to those who can.

Stage 5: Proposals and tenders

Most projects will go through a proposal or tender stage. This is not just submitting a price. Often there will be an element of pre-qualification, writing compelling method statements, making a presentation, negotiation and contractual points. During this process, the client will be looking for signs that he can work with you. If you get this stage right as well as the previous stages, you may well be able to secure the project with a better financial package because the client is happy to pay a little more for the right person. That's why all the initial work is so important.

You will probably stand less chance tendering for a project where you have had no previous client contact compared to a project where you have been building up a rapport. Unless you are happy to be judged on price alone, pick carefully from the opportunities that come your way.

If you have worked through the previous stages, then, by the time the project comes to tender, you will have a good idea what the client wants from his professional services team. Many tenders do not explain the background to the project and many tenders don't have an evaluation criteria established to assist selection – especially at the pre-qualification stage.

A client will be more likely to short-list your firm at the pre-qualification stage if he knows your firm. If he is comfortable with your firm winning, then you stand a good chance of getting through to the final tender stage. The unknown firm will find it difficult to win through to the final stage if there are numerous well-established firms also chasing the work.

This stage also deals with the art of pitching for work. How to plan the pitch and prepare the team for presentations? Looking at the content and delivery of the pitch is one of the keys to success.

Stage 6: Delivering added value and obtaining repeat business

If you deliver a good service then you stand a good chance of getting repeat business. So often professionals are so keen to win new clients that they ignore existing or past clients. They let other professionals to sneak in and steal the work from under their noses.

From the very first day of your new assignment, you will need
to put in place a strategy for securing additional work and keeping
your competitors out. Within this stage we explore the benefits of
preparing a project client plan, which builds upon the client capture
plan developed in earlier stages.

You will need to put in place client account management and
making sure that all the time the client makes contact with your firm
(the 'service touches') they are good experiences.

There will be a need to explore cross-selling opportunities, creating
new relationships within the client organisation and keeping on top of
providing a good service. Feedback is important and establishing the
level of client satisfaction is critical to enable your firm to continuously
improve, and thereby enhance, the opportunities for receiving repeat
work.

Stage 7: Building credibility
Although this is Stage 7, it should be happening all the time. At
every opportunity the professional should be building up a bank of
material to help secure future work. This would include case studies,
references, producing news items and articles to build up the firm's
reputation. Clients feel at risk when dealing with professionals with
no apparent track record or list of satisfied customers.

To help you through all the first six stages and with new opportunities, you need to continuously examine what you have been doing
and recording information. Also build up references and encourage
clients to write in with good endorsements. In fact, keep everything
that may be useful to secure future work.

Worked examples and gender

I have provided many sample letters, figures and tables to illustrate
the various techniques used within the seven stages. The readers may
wish to use these to create their own to suit their own individual
circumstances. To make the examples more realistic, I have used a
fictitious firm of architects called Wren & Barry Ltd and I have also
provided them with a fictitious list of clients.

To simplify the text I have chosen to use the male gender throughout.

Stage 1: Selecting the clients you want to work for

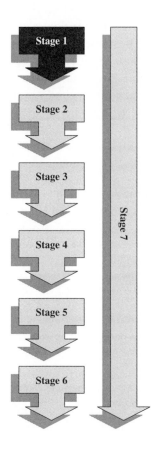

1.1   Your client portfolio

1.2   A strategy

1.3   Effort versus reward

1.4   Lifetime value of clients

1.5   Picking your moment

1.6   Creating a pipeline

Stage 1: Selecting the clients you want to work for

1.1 Your client portfolio

Many books on selling professional services only concentrate on targeting new opportunities. This would ignore the easiest work to win, which is from existing and past clients. Assuming you have provided a good service and your current and past clients still have work to commission, then this should be your main priority.

Obviously you still need to, at the same time, target new opportunities so that you are gradually building up your portfolio and replenishing where current clients have come to the end of their commissioning period.

Examine current portfolio

Before deciding a campaign of action to secure more work, it will be useful to examine your current and past work so that you can get an indication of the following:

❑ Who were your best and worst clients, and why?

❑ What kind of work was most profitable?

❑ What sectors gave you the most work?

❑ What skills were being commissioned?

It makes sense to focus most marketing and sales efforts towards those clients who are going to continue to give you profitable work. There is little point in spending a lot of time and effort trying to 'win' work from clients who do not value your work, pay below the market rate and have no interest in a long-term relationship.

So the first task is to divide your current and past clients into three categories, which we will refer to as gold, silver and bronze. **Fig 1.1**

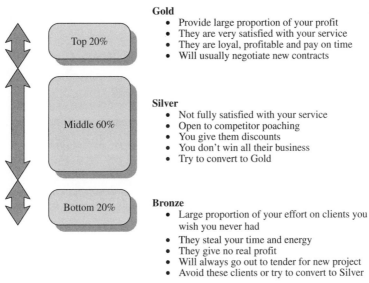

Gold
- Provide large proportion of your profit
- They are very satisfied with your service
- They are loyal, profitable and pay on time
- Will usually negotiate new contracts

Silver
- Not fully satisfied with your service
- Open to competitor poaching
- You give them discounts
- You don't win all their business
- Try to convert to Gold

Bronze
- Large proportion of your effort on clients you wish you never had
- They steal your time and energy
- They give no real profit
- Will always go out to tender for new project
- Avoid these clients or try to convert to Silver

Figure 1.1　Gold, silver, and bronze client criteria for a typical client portfolio.

shows the criteria for each category. If you are a large organisation then this can be carried out for each office, each department or each major skill set. You may find this quite revealing because you just might find that

❑ the clients who give you the most turnover might not be gold clients but could turn out to be bronze;

❑ the same clients may appear as gold in one office or department and a bronze or silver in another office or department;

❑ large blue chip clients might be more demanding and less profitable than smaller not so well-known clients.

Take this analysis one step further and work out what the total amount of fee income is for each category and the proportion of clients falling within each category.

Table 1.1 shows the top clients, by fee income, for 1 year for a fictitious firm of architects called Wren & Barry. The table shows that nearly 80% of the fee income is generated by just six clients. These

Table 1.1 Top clients by annual fee income

Top clients by fee income for Wren & Barr

Rank	Client	Sector I:Industry C:Commercial L:Leisure H:Health	G:Gold S:Silver B:Bronze	Annual fee income	% of total	Cumulative total	Cumulative %
1	Brown & Jones	C	G	188k	25.1	188	25.1
2	City Builders	I	B	135k	18.0	323	43.1
3	Midshire Leisure	L	G	90k	12.0	413	55.1
4	Barker & Conlan	C	G	88k	11.7	501	66.8
5	Allan & Parker	C	S	45k	6.0	546	72.8
6	Healthlands Developments Ltd	I	B	40k	5.3	586	78.1
7	SE Health Trust	H	B	32k	4.3	618	82.4
8	Drake Pearson	C	S	27k	3.6	645	86.0
9	Spruce & Willow	C	B	12k	1.6	657	87.6
10	Crow Machines Ltd	I	B	9k	1.2	666	88.8
11	Bells & Whistles	I	B	6k	0.8	672	89.6
12	Drake Pearson	C	B	4k	0.5	676	90.1
13	Others (20 No)		G(2No) S(6No) B(12No)	7k 22k 45k	0.9 3.0 6.0	750	100
	Total			750k	100		

4

six clients account for almost 20% of the total number of clients that the firm is currently working for and three are classified as gold clients.

If you were to repeat this exercise every year, you would see how different clients may feature in different years. This emphasises that you need a good spread of clients to give a continuous stream of income and not be reliant on just a few good clients. This also underlines the need to add to your client portfolio, which we will be looking at later.

Also in **Table 1.1** we can examine the mix of gold, silver and bronze clients within the portfolio and how much fee each client type contributes, again using our fictitious firm Wren & Barry. **Fig 1.2** shows the analysis of client and fee income for the firm.

From experience, you may find that your results might follow the Pareto rule, which states that for many events 80% of the results come from 20% of the outcomes. The rule was named after an Italian economist Vilfredo Pareto, who discovered that 80% of the income in Italy went to 20% of the population. The rule has now found a home within business and it is often said that

❏ of your sales, 80% comes from 20% of your clients, and

❏ the least profitable 20% of your clients will probably take up 80% of your spare management time as they will be more troublesome and demanding. These will also be typically bronze clients.

Gold	Silver	Bronze
16%	25%	59%

Proportion of gold, silver and bronze clients

Gold	Silver	Bronze
49%	14%	37%

Percentage of total fee income from gold, silver and bronze clients

Figure 1.2　　Analysis of clients and fee income for Wren & Barry.

Therefore, in an ideal world, you want to increase your share of gold clients. This can be achieved by converting the bronze to silver and the silver to gold. You will probably find that many of the bronze clients have approached you, perhaps having previously employed your competitors and fallen out with them. There is a good chance that your gold clients were clients you had previously targeted or who came to you through recommendation. Gold and silver clients tend to value quality of service where bronze clients tend to be focused mainly on the cost of delivering that service.

In our Wren & Barry example, we can see the following from **Table 1.1** and **Fig 1.2**:

❑ Of the fee income, 49% comes from the gold clients who account for only 16% of the client portfolio.

❑ Over 80% of the fee income comes from seven clients (just over 20% of the total number of clients).

❑ From the top 12 clients, who account for over 90% of the fees, we see that the commercial sector brings in 48.5% of the total fees, and industry 25.3%, leisure 12% and health 4.3%. This is useful to know because, in this example, Wren & Barry have been spending more time pursuing, say, health in recent years, yet the results were not reflecting this. Obviously, we need to be aware that there is a time lag between the marketing and sales activity and the work being secured from that effort.

It is fundamental that you know how your current portfolio is made up. Going through the same investigation as carried out by the Wren & Barry example will give you the knowledge to allow you to move forward with more confidence in deciding how to pursue more work.

Saying goodbye to clients

The phrase 'the customer is always right' does not hold within the world of the professional services firm if the client is

❑ awkward and too difficult;

❑ unduly complaining;

❑ taking up far too much time to manage relative to the fee.

Then be brave enough to make the decision to part with these clients, rather than take on new projects for them. For these clients are, according to our criteria, 'bronze' clients.

Having carried out this analysis, for your own business, you have been able to categorise those clients that are a drain on your resources. You do not want bronze clients within your portfolio unless they pay for the privilege. Therefore, the following are the outcomes of this exercise:

❑ Invest time and effort to convert bronze to silver and in time to gold by making them appreciate the service delivery rather than just the cost of that delivery. You can do this by building strong relationships. This is discussed in more detail in Stage 6.

❑ If you know that the bronze clients are not convertible, then raise your fees to service them. Let them pay more for taking up a disproportionate percentage of your management time. If by doing this they don't come back to you, then that's fine because hopefully they will clog up your competitor's portfolio and take up their time and effort.

Number of clients

The breadth and mix of your client portfolio is as important as the quality of each client. Ideally, you need to have a variety of client size contributing to your turnover. It would not be good business sense to be reliant on just one or two key clients because they may suddenly stop giving you work for a whole number of reasons, thereby leaving you very vulnerable.

Ideally, your client portfolio should have a spread where

❑ no client contributes more than 25% of your turnover, preferably no more than 15%;

❑ perhaps one or two clients would each contribute approximately 15% of your turnover;

❑ then the next band of say four or five clients contribute between 7.5 and 12.5%;

❑ the remaining clients contribute the rest.

You may decide a different mix more suitable to your needs depending on your organisation size and potential number of clients in the market place that may need your services. Also, you need to bear in mind that during the life of the client relationship, a client may appear as one of your key clients in one year and in another year as a minor client. What is important is that you are aware of the lifetime value of your client (see Section 1.4) and keep in touch to receive all their future work.

Sector exposure

Just as it's leaving yourself vulnerable working for one client, you would also be exposing yourself to risk by just working in one sector. Some organisations, for example, will only work in the residential sector. In this situation, there only needs to be an economic downturn and the sector slows right down. If possible, it would be better to have work from the following:

❑ A mix of public and private sectors (perhaps education and health as well as industrial and commercial).

❑ Sectors that are perhaps more resistant to economic downturns.

❑ Some sectors that are more profitable than others, especially if you have specialist skills. Working for the pharmaceutical sector, for example, might be more demanding and profitable than the warehousing sector.

Looking at diversification, it would be preferable to look at sectors that you are not currently working for, but which would require the same skill set. Perhaps an organisation specialising in the health sector, which is highly serviced and complex, could capture work in the pharmaceutical sector. Equally, perhaps, those serving the housing sector could move into residential care homes where the type of building is usually of domestic scale and construction.

Who are your competitors and who do they work for?

When putting together a strategy for obtaining more work, you may get some insight into what to pursue by looking at your competitors. Take a look at your competitors and consider the following:

❏ In what sectors do they compete?

❏ What size of contract do they undertake?

❏ Are they more successful than you are and if so why?

❏ What client portfolio do they have? Can you learn anything from how they operate to secure work?

Table 1.2 shows a competitor analysis for Wren & Barry. From this example, you can see that

❏ most of the competitors have local and regional coverage with only a few with national coverage;

❏ some only work in one or two sectors;

❏ some competitors from outside the area are focusing locally.

From your research into your competitors, you may uncover that one or two are going through a bad patch with high staff turnover and losing clients. If this is the case, you might decide to target some of their remaining clients in the hope that they might be right for a switch to other professional service providers.

When examining your competitors, ask a colleague (outside your firm) to telephone the competitor and ask for a brochure. Ask your colleague to report back on the following:

❏ How was the request handled? Was your colleague put through to the right person?

❏ What questions were they asked? Were they just asked for their name and address or were they also asked if they had a project in mind?

❏ Did they receive the brochures promptly? Were they what had been requested?

❏ Was there an accompanying letter or any additional information?

Table 1.2 Competitor analysis for Wren & Barry

Competitor analysis for Wren & Barry

Competitor	Coverage			Sector activity						Comments
	Local	Regional	National	Commercial	Industry	Health	Residential	Leisure	Education	
Latch & Phelps	X			X				X	X	Well established with good local and regional clients
Industrial Design Ltd		X		X	X					Only work within the industrial sector
Barrow, Jones & Bond	X	X	X	X	X	X	X		X	Large national business with good blue chip clients
Barlow Studio	X			X						Mainly interiors and space planning. Have taken on Education specialist
Clarke, White & Smith		X	X							Haven't worked local yet but have been pitching for work locally
Chris Pen Associates	X	X		X			X			Only one partner, have succession problems and associates leaving
Carter & Lynch Ltd	X	X		X	X					Very good designers and have won many design awards
RQT Design	X				X					Very competitive business and seem to be winning everything locally

❏ Did the competitor follow up with a telephone call to enquire if they could be of any further assistance?

When I have carried out this exercise in the past, I have always been amazed that professional firms do not follow up. It appears that there is no link between the marketing department (sending brochures out) and the business development/sales team to chase the possible lead. Needless to say that your own business would need to look at how it handles requests of a similar nature. This is discussed further in Stage 7.

1.2   A strategy

Many professionals are quite happy to receive enquiries from any potential client. A client might just telephone out of the blue and request a meeting and a proposal for a particular project. This might seem fine, and on occasion this will be, especially if a previous client has sung your praises and you have a potential client who is in buying mode.

However, there are some potential dangers in taking on an enquiry without some initial investigation. Unless the potential client has approached you because of your reputation, the likely selection criteria will be based mainly on price. Price is fine if all those being considered for the project are of similar skill, size etc. In which case, other factors may come into the selection process.

There are potential problems in taking on board an enquiry from a client who is not known to you. Because you haven't gone through the targeting and relationship-building process, explained later in this book, you don't know if

❏ the client is serious in appointing you or is just fishing for ideas;

❏ the client might have retained consultants and might be looking for a comparative check price;

❏ the person contacting you has the relevant authority to appoint you;

❏ there is finance in place for the project;

❑ the client decision process has properly commenced and at what stage the enquiry is.

This doesn't mean that you should turn down the enquiries that come in out of the blue. They may well turn out to be worthwhile jobs and you could go on to have a long profitable relationship with the client undertaking numerous assignments. It is, however, worth having an initial checklist of questions ready so that you can quickly get an overview of the status of the potential project. **Table 1.3** is a sample checklist which you may find useful as a starting point. You could ask some of the questions during the first discussion and others during the briefing stage. The most important thing is that you undertake the enquiry knowing where you stand and are able to roughly evaluate the probability of success.

It is not good business sense to rely totally on enquiries coming in and being just reactive. It makes sense to have a plan and be proactive in winning work. You should decide on whom you want to work for and the type of work you are going to undertake.

Having looked at

❑ your current client mix,

❑ your potential vulnerability in respect of sector or client dominance,

❑ what your competitors are doing,

you can begin to put together a strategy to determine what kind of client you want to secure work from.

Prioritise your effort

Unless you are just starting out with no track record, and no clients, you need to prioritise your efforts in securing additional work. You need to maximise every hour you spend.

Therefore the following would be a good starting point to prioritise your efforts (see also **Fig 1.3**).

Priority one: existing clients
Good existing clients are the best starting point for additional marketing. Obviously, this would not apply to clients where the relationship

Table 1.3   **An initial checklist to evaluate enquiries**

Target the best opportunities A checklist to evaluate enquiries
Client information
Who is the ultimate client?
Are we working subcontract?
Who will pay our fee?
Has the client any consultant advisors? Who are they?
Is the client knowledgeable about construction?
Is the client known to us?
Have we worked with this client before?
Does the client contact have authority to appoint us?
Does the client have previous construction experience or working with consultants?
The competition
Are we in competition?
How many others?
Are we competing against similar organisations?
Is any speculative work required to win the project?
What will be the cost of speculation work including expenses?
Is there a deadline?
Funding and fee
Is funding for the project and fees in place?
What is the estimated contract value?
Who prepared the estimate?
Is it realistic?
Is cash flow linked to financial year etc.?
What fee do we expect?
Is a fee agreement in place?
Risk
Is the programme feasible?
Are appropriate resources available?
Does the project carry any particular risk?
What is the percentage chance of securing the project?
Given the answers to the above what chance do we have?
Probability increases if:
❏ The project is for an existing client
❏ The project is in an existing sector of expertise
❏ We are liaising direct with the client
Probability decreases if:
❏ We are responding to an advert
❏ The client hasn't worked with us before
❏ We are subcontracted to someone else

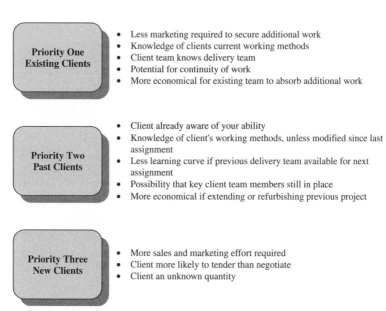

Figure 1.3   Strategy for pursuing additional work.

is not good. Hopefully, you have already established a good working relationship so a great part of the marketing effort has been done. Therefore, you need to quickly establish the following:

❑ Does your client contact have any more work to hand out?

❑ Can your client contact introduce you to any other people in the client organisation that may be giving out work?

❑ Can your client contact introduce you to other organisations that may be giving out work?

Once you have established that there is potential work from one of these three areas, then maintain contact or build relationships so that when there is work being commissioned you are the natural choice or at least put on the short list. Stage 6 covers in more detail how to secure work from existing clients.

Priority two: past clients
How often have you seen people do a good job and then years later, someone else comes along at the right time and grabs the next job? This

happens all too often, especially when the initial people have moved on to seek pastures new and not kept in touch. This might also happen if the initial client contact has moved on to another organisation or your own internal key account manager has left your employment. Their new employer might be now working for your past client!

With past clients it should be relatively easy to make contact again and arrange a meeting and restart building up a relationship. Never lose contact with past clients. Even if they have come to the end of their building programme, they can still be useful. They can

❑ give you referrals and references;

❑ allow your potential clients to visit their site to see your work;

❑ move on and you could follow them and get into a brand new client and even a new sector.

Priority three: new clients

Many people who are looking to attract more work will usually target new clients. Developing new clients should be considered alongside obtaining work from past or existing clients. It's good to have a healthy mix of repeat work and new work. Repeat work will usually be more profitable because of the following:

❑ There is no or little marketing expenditure required.

❑ You know how the client operates and his systems.

❑ Usually, it will not be for a low fee because the client values your ability. However, a past or current client might wish to, or be obliged to, still go out to tender. In this case, hopefully the client will pick similar organisations; so experience, skill set and service delivery will be taken into account as part of the selection criteria.

Securing work from new clients is the hardest of the three priorities, but can be very rewarding. Most large professional firms will have their business development managers or directors targeting new clients, leaving the delivery teams to secure the repeat work. Although this seems, on the face of it, a sensible use of resources, they do not ignore, or short change, the ongoing building of relationships (and maintaining them) with existing and past clients.

Selecting your targets

Unless you are providing a very specialistic service for a very limited pool of clients, there is a need to take some time in selecting your target list. The following are points to consider:

❑ **Location:** Are you able to provide a service economically from a distance, especially if you are based in one location? Unless your service is highly specialistic and sought after, it might be better first to restrict and then to select your targets from a local area.

❑ **Sectors:** Are you providing a service within a specific sector, e.g., commercial, retail, industrial? Will you have sufficient track record to get the target client's interest within their sector? Of course, you might be able to transfer skills from one sector to another.

❑ **Size of client:** A large client will probably give large projects to other large organisations. The client will want to be sure that there will not be a resource problem. However, large client organisations will also have small projects to hand out. So, if you are a relatively small business, there is no need to exclude large organisations.

From this, you can create selection criteria and then use that to create a target list of potential clients. On the basis of these criteria you can create a list of targets by doing the following:

❑ **Purchasing a mailing list that is already available.** The more specific your requirements are the better the quality of the list.

❑ **Researching your own list.** You can buy various sources of data (e.g. Kompass), which allows you to search on the basis of your location, sector, and turnover criteria.

❑ **Commissioning your own target list.** Organisations such as the local Chamber of Commerce may have a service that can provide you with a target list based on your criteria. The list can break down the contacts depending on whatever criteria you want. A good starting point, once the geographical spread is decided, would be the size of target client based on the number of its employees.

A selection criteria based on turnover might be a consideration, but within the building industry it is usually the head count that gives a better indication on the spend on buildings. The added benefit of buying a list, which is in an electronic format, is that it will probably lend itself to populate your database or client relationship management (CRM) programme, thereby saving many hours of input time.

The need to comply with the Data Protection Act

You will need to take expert legal advice to make sure that the way you handle data about current, past and potential clients complies with the current legislation such as the Data Protection Act (DPA), regulations and codes of practice. Generally you will need to ensure the following:

❑ Your data is secure and does not fall into the wrong hands.

❑ Data should not be stored longer than is necessary.

❑ The data is up to date. If a potential client wants his data removed you must do so.

❑ Keep contacts up to date on how you use their data and give them an option to be excluded.

❑ When buying lists you need to check with the suppliers that the data has been compiled in a proper and legal way.

Sector penetration

When you prepare your target list you may decide that there is scope for work from a particular sector. This might be because

❑ you have just completed several projects within a sector and believe you have established an expertise;

❑ you are aware that due to the current economic climate the sector will be growing while other sectors in which you are currently active will be diminishing;

❏ you have developed experience in one sector and you believe that this experience and skill set can be transferred to other sectors.

When going through the process of securing new clients (priority three), it is easier to prepare on a sector by sector basis. This has the benefit of getting into the mindset of your target clients and their business needs. When you contact a potential client, the feedback you obtain about the sector can be used straight away in your discussions with the next one.

In preparation, you will need to establish your track record and be able to quote facts and figures when talking or communicating to target clients. If your sector is large and there are subsets to the sector, then experience in one area might not be appropriate in another; for example,

❏ experience in pre-prep schools might not be seen as experience for university work (both education sectors);

❏ experience in private fitness centres might not be appropriate when pursuing local authority leisure centres with large swimming pools (both leisure sectors).

However,

❏ sports halls for local authority leisure centres would be good experience for sports halls on a university campus;

❏ laboratory experience within the pharmaceutical sector might be seen as good experience for laboratories in hospitals or universities.

When considering sector penetration, take time to consider your current position. In **Fig 1.4** the current sector penetration, based on fee value within 1 year for Wren & Barry, has been identified. Also, within this example, the target growth has been highlighted, with a target growth in the education sector seen as the main priority (+15% of total current annual income coming from education). This example also shows that Wren & Barry want to increase turnover by 43% over a set period of time, say 3 years. This gives a projected growth in the order of 15% per year. In this example, it can be seen that no growth

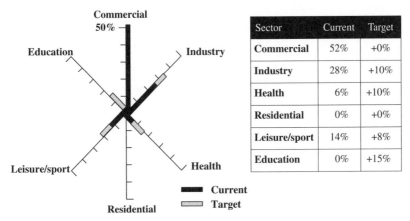

Sector	Current	Target
Commercial	52%	+0%
Industry	28%	+10%
Health	6%	+10%
Residential	0%	+0%
Leisure/sport	14%	+8%
Education	0%	+15%

Figure 1.4 Current and target sector split for Wren & Barry.

is targeted for the commercial sector. This is because it is appreciated that they have been very strong in this sector and holding onto their share of the sector would still require much proactive work.

There are several points to consider when looking at new sector penetration. These are outlined in **Fig 1.5**.

1.3   Effort versus reward

It can take several years to build up a relationship with a new client. This is very time-consuming and costly. You need to consider the lifetime income from the client. Therefore, client types are of interest here.

Public sector

More than 25% of the UK workforce is employed by the public sector. This requires a huge annual spend on new facilities and the refurbishing of existing accommodation. No matter how large or small your business, there are always opportunities to secure work from the public sector. The perception seems to be that only large firms can afford to go through the public sector procurement processes. This is not so. In 2004/2005, small and medium-sized business secured 22% of central government contracts and the majority (59%) of the

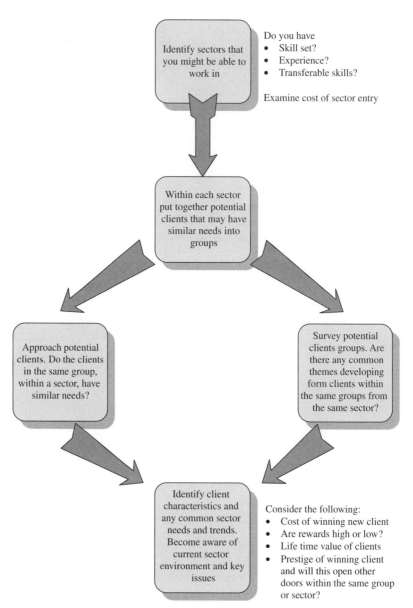

Figure 1.5   Initial considerations for new sector penetration.

total value of the local authority projects. Looking across the whole European Union, 16% of all businesses that were awarded public sector contracts employed 10 or less employees.

If you prioritise your efforts and target projects that you are able to deliver then you may find the public sector a fruitful market place. The public sector consists of the following:

❑ Central government and agencies

❑ The Scottish Executive, National Assembly for Wales and the Northern Ireland Assembly

❑ Ministry of Defence

❑ Local authorities

❑ The National Health Service and the numerous local trusts

❑ Universities and colleges

Compared to some private sector clients there are distinct advantages in working for the public sector. These include the following:

❑ The 'client' is usually longstanding, stable and pays within 30 days or any other agreed period.

❑ The selection procedure is open and usually based on experience, ability to deliver the project and value for money.

❑ Funding, or a budget, is usually in place so there is less chance of the project being aborted.

The main downside is that most projects are advertised and there are many professional services firms chasing each project. Therefore, as discussed earlier, it is important to target the specific areas you want to work for and make yourself known to the client before projects are advertised. On many projects there are professional services firms already 'attached' to a project, having carried out feasibility studies, etc. and so have a distinct advantage to secure the follow-on work. In these cases it is important to try to win small assignments, which may fall below the threshold, requiring the project to be advertised.

It might make business sense to undertake small assignments at a reduced fee (as a marketing exercise) in the hope of standing a better chance of winning the implementation project.

Many public sector organisations, especially local authorities, have approved supplier lists for certain types of work. These approved lists are usually used for the lower value contracts below the value set in the EU procurement directives.

These EU directives require all public sector contracts for buying goods and services within the EU above certain thresholds to be advertised. The contracts are awarded according to principles that ensure non-discrimination, equal treatment and openness. These projects are advertised in the following:

❑ Official Journal of the European Union (OJEU, previously called OJEC). This publishes most contracts, on a daily basis, that are above the value set in the current European Directive. You can receive this on line through TED (Tenders Electronic Daily) or through many alert services that filter opportunities based on your interests. These include Supply2.gov, Business Link or through many private commercial organisations that service the market. The stationary office can provide you with OJEU on CD-ROM through subscription and also has a service whereby scanned extracts can be sent to you electronically.

❑ Advertisements in trade journals or regional newspapers.

❑ Supply2.gov.uk internet site that can help you search for lower value contracts.

You can also obtain information from Euro Info centres (EIC) that have been set up thoughout the United Kingdom.

Many of the larger high-value government contracts will be won by large companies. This still provides opportunities for the smaller professional services firm or smaller contractors securing work as subcontractors. Many projects come out into the market place on a design and build procurement basis. Designers should therefore forge links with contractors to secure work, which they could not hope to win on their own account. Some projects that come out as public private partnerships (PPP), private finance initiatives (PFI) or as part of initiatives such as building schools for the future (BSF) can be very large and therefore prohibitatively expensive for smaller businesses to take part in the tendering process.

Private sector

The private sector is very different from the public sector. Clients have more flexibility on how they decide to procure projects. They might decide to negotiate direct with one professional services provider or ask several firms to submit proposals or tenders. Also, because public money is not being used, the private sector client can use whatever selection criteria they wish (subject, of course, to certain laws concerning age or race discrimination, etc.). Unlike the public sector, where the criteria has to be declared and feedback given to unsuccessful bidders, no such requirements are placed on the private client, although it is good practice and provided by enlightened clients.

When considering whom to target within the private sector, it is worth considering the long-term potential and lifetime value of target clients. Therefore, some research into their building portfolio and long medium to long-term plans would be beneficial.

Single-property clients

These are probably the ones of least priority. It is fine if you become aware that work is going to be commissioned on the single property, but rewards might be less related to effort when considering other opportunities. Obviously, if the project you have identified is a sizable one and is sufficient reward in itself to chase, then there is nothing wrong in pursuing it. If, however, the project is much smaller than you usually tackle and does not seem profitable on its own, then you may think twice about taking it on or spending too much time pursuing it.

Single-site, multi-property clients

The single-site client who has a campus made up of many buildings is one of the best opportunities to chase; this is because of the following:

❏ All projects will be within your geographical catchment.

❏ Once appointed and working on campus, then you could become the local expert as you get to know more about the site. You will get to know about its services, planning restrictions and other constraints. As long as you continue to deliver a good service, you are in a strong position to continue on-site dealing with projects as they arise.

❑ You will have a better opportunity to keep the client relationships warm by undertaking the smaller projects in between the larger assignments.

Multi-site organisations

These are obviously of significant interest and by their very nature will be the ones that your competitors will also be chasing. Clients such as supermarket chains or other multi-site businesses would fall into this category. These clients suit multi-office professional firms who can offer to service all their sites with local teams. However, many of these clients do not want to restrict themselves to the larger firms providing professional services and so may split the country into regions and have firms competing against each other in one or more regions.

There could be some problems in servicing multi-site organisations, such as the following:

❑ There might be different commissioning people within the client organisation for each site or region, which increases the selling effort and building relationships on various fronts.

❑ If you are a single office firm, then you might find it difficult to service all the sites economically.

Contractors and developers

These organisations would have their own business development teams keeping track of new activity within the construction arena. So by courting these clients (first tier), you could become a second tier, or subcontractor. This would reduce your marketing effort. Many clients are now looking for a total turnkey solution, so being part of a large team can still give you additional workload.

Other consultants

Some consultants are useful in that they are happy to make introductions to their clients. Networking with these consultants is discussed in Stage 4. Some consultants become commissioning agents on behalf of clients and so can be targeted as potential clients in their own right. Project management companies fall within this category.

1.4   Lifetime value of clients

The lifetime value of a client is an important consideration. If the client is able to give you many projects on a regular basis then the benefits are as follows:

❑ Initial marketing costs spread over many projects so there is less spend on a project by project basis.

❑ You have a learning curve initially, but thereafter, you know how the client wants to work and you get to know his business.

❑ You are able to retain a team that is dedicated to that client. This saves on the cost and management of recruitment.

❑ It gives more stability to your business, and with several clients giving regular work it boosts the value of your business and brand.

❑ It helps to recruit better people who are looking for stability and progress in their careers.

In return, many clients who have multiple projects know their value to the service providers and in return can

❑ negotiate lower fees and rates for volume work;

❑ insist that they have dedicated teams to service their projects;

❑ create framework agreements to enable them to have choice and not be restricted to one supplier.

Given this situation, it is also difficult to win these clients. These clients are sophisticated and know that there is a need to consider new service providers. By doing your research, you can find out who the current providers (your competitors) are, what the arrangement for the provision of services is and when the renewal time is. Many of these framework deals last for 3, 5 or more years. So you don't start making your pitch in the year of the change or retendering. You need to plan ahead, perhaps making inroads some 2–3 years earlier. This might sound a long courtship and that's why many service providers don't make the necessary early commitment. There is always the temptation of something else around the corner that could give immediate results. Ideally, you need a mix in your portfolio of target clients so that you have some long-, medium- and short-term targets.

When looking at the reward versus effort relationship, you will be able to make some quick calculations on potential lifetime earning capacity for potential clients. Don't worry about being extremely accurate; what you need initially is some indication of key facts. The example in **Table 1.4** shows a simple format for the lifetime values of clients for Wren & Barry. From this table you can see the following:

❑ The current best client (Brown & Jones) has no more future work or fees due.

❑ The design and build contractors or developer clients (City Builders and Heathlands Developments Ltd.) are the best current lifetime value clients. On the basis of past performance, it can be assumed that regular work would be forthcoming from these clients for many years to come.

❑ It is always good to have some regular smaller projects (from Crow Machines Ltd. in the example) as this provides good cash flow and can be introduced to a team currently engaged on large projects as workload diminishes.

To assist in your calculations do some initial research, perhaps call the buying department within the target client organisation and ask some of the following questions:

❑ What is the current procurement method?

❑ When is the renewal time?

❑ What will be the format of reselection? Will it be tender on rates, a specific project?

Table 1.4   Lifetime value of current Wren & Barry clients

Lifetime value of current Wren & Barry clients

Current rank	Revised rank	Client	Current annual fee	Year 2	Year 3	Year 4	Year 5	Year 6	Year 7	Year 8	Year 9	Year 10	Total
1	5	Brown & Jones	188k										188k
2	1	City Builders	135k	50k		75k		75k		75k		75k	485k
3	4	Midshire Leisure	90k				100k						190k
4	6	Barker & Conlan	88k										88k
5	9	Allan & Parker	45k										45k
6	2	Heathlands Developments Ltd	40k		50k		50k		50k		50k		240k
7	10	SE Health Trust	32k										32k
8	11	Drake Pearson	27k										27k
9	12	Spruce & Willow	12k										12k
10	8	Crow Machines Ltd	9k	4k	5k	5k	5k	5k	5k	5k	5k	5k	53k
11	3	Bells & Whistles	6k	30k		200k							236k
12	7	Drake Pearson	4k	75k									79k

❏ What will be the selection criteria?

❏ How long have current suppliers been providing services?

❏ When was the last time a new supplier was introduced to the team?

❏ What are the future spending plans?

Some clients, especially in the private sector, might not be so forthcoming with this information, in which case, some research on the internet and a look at the annual company report to shareholders could be informative.

Obviously you need to stay clear of clients who are in the decay stage of their construction lifecycle, as initial costs of winning the work might not be recovered in profit for just the remaining project portfolio.

The other point to consider is the match between supplier and the work being commissioned. Perhaps during the early years, there may be a large investment on big projects, which will only be suited to the larger service providers. There might be a point when all the big projects are complete and what remains are the smaller projects and extensions. These may not be of interest to the initial larger providers but would be ideal for the medium-to-small professional service firms.

Also, look at your competitor websites and see how much work they are doing for your target clients. If there are several of your competitors serving that client, then see if they are on a framework scheme and if the framework is grouped according to project value or regions.

The key message here is research, as much as time and resource permits, and is realistic on what may be winnable. Once you have established the target list, then make time and take the effort to service that list so that you can develop a pipeline of potential business.

When doing the research make sure that you try to target the most profitable and, if possible, long-term clients. You will need to investigate potential clients from the following view points:

❏ Which is doing well or who might be doing well? This could be influenced by market conditions, change in the law or the client suddenly having a product that is doing well and expansion looks likely;

❏ Their sector or markets. An upturn in the economy might see tourist-related business doing well and a downturn could see residential clients or those serving that market doing badly.

Your existing client base and selling efforts might be satisfactory for the next 3–5 years, but you need to replenish the existing customer base continuously. This is especially the case if you have been totally reactive and have not had a proper sales campaign in place. Take some time and review the average life of your customers. Is it just one project? Have you been ignoring existing clients and have failed to pick up their next project?

1.5   Picking your moment

Having established selection criteria to create a long list of potential targets, you need to start making some enquiries to see if they are in buying mode.

Apart from large organisations, such as the large retail outlets, most potential clients might be in between projects. This period could be months or several years, so there needs to be some methodology to determine when they will be entering a buying mode. The change from not being in the market to needing assistance is marked by a switch point. The good sales people will recognise what switch points will apply to their potential clients and even better sales people will be tracking potential clients on their target list, just waiting for a switch point to occur.

Switch points are points in time when the client switches from 'off' (not buying) to 'on' (buying). These switch points can be activated by the following:

❏ New incoming high-ranking person. A new operations director or chief executive might want to make changes. These usually occur after they have had sufficient time to evaluate the business and prepare an internal strategy for change.

❑ A point in time after exceedingly good business results when the company needs to expand to continue growth, or after a period of decline and a need to downsize and possibly sell assets, which could be buildings and land.

❑ The incumbent service provider fails and needs to be replaced.

❑ A term contract or framework agreement comes to an end.

❑ An acquisition by another business or the target client is in acquisition mode. During acquisitions there is usually a rationalisation of facilities, which often creates some sort of building work.

❑ A change in technology. Look at what happened to the newspaper industry with the introduction of new technology.

The best way to determine the onset of a switch point is to talk to the target clients. Talk to the purchasing, property, estates departments. Start to build up a rapport and find out facts. A typical telephone conversation could be as follows:

Mr Barry: *Hello, I wonder if you could help me? Could you confirm that Mr Phillips is in charge of appointing consultants such as architects?*

Receptionist: *Mr Phillips is our operations director, I am not sure if he appoints architects. Do you want me to put you through to Mr Phillips?*

Mr Barry: *Yes please, does he have a PA?*

Receptionist: *Yes he does, a Mrs Jones, I will put you through.*

Mr Phillips: *Hello, Mr Phillips here.*

Mr Barry: *Hello Mr Phillips, I wonder if you could help me? I am Mr Barry from Wren & Barry architects. We are based only a few miles from your site and I am calling to enquire if there is any work coming up in the near future we could perhaps help you with?*

Mr Phillips: *Well, we have just finished a new warehouse about a year ago.*

Mr Barry: *Oh that's a shame. Who were your architects on that project?*

Mr Phillips: *Industrial Architects Ltd, a local firm.*

Mr Barry: *Yes I know them. Is there anything else coming up?*

Mr Phillips: *Yes there is, but it's all confidential at this stage.*

Mr Barry: *I totally appreciate that. We work for many clients in a similar sector to yours and fully understand the need for commercial confidentiality. I wonder if you could let me know how far off the project is?*

Mr Phillips: *It all depends on the main board in Paris, but perhaps in about a year's time.*

Mr Barry: *I will make a note of that. Would it be OK with you to call again in say 6 months and see how that project is progressing?*

Mr Phillips: *Yes you can.*

Mr Barry: *Thank you. To help me when I call again can you give me a project name or something to describe the project so that when I call in 6 months time I can enquire how it is progressing?*

Mr Phillips: *We refer to it as Project Green Park.*

Mr Barry: *Excellent. One final question if that's OK with you. How do you go about selecting consultants such as architects?*

Mr Phillips: *That's all done by our procurement team who are based here. I suggest that you contact the head of that team, a Mr Andrews.*

Mr Barry: *Thanks very much for your time. I will speak to Mr Andrews and give you a call in, say, 6 months.*

From the above conversation, Mr Barry was able to establish the following:

❏ Mr Phillips is the operations director and has a PA called Mrs Jones (the 'gatekeeper'). When Mrs Jones is not about, he does pick up the phone himself.

❏ They just completed a warehouse and used a local firm of architects.

❏ Mr Phillips does know what work is coming up and has identified a project, which can be put on the leads list.

❏ Mr Phillips might be an influencer in consultant selection, but it appears that the procurement department is in charge of the process and the key contact appears to be Mr Andrews.

❏ The board, based in Paris, has the final say.

As a follow up to this call, Mr Barry, the architect, should do the following:

❏ Contact Mr Andrews of the procurement team and establish any requirements for being considered for work. A good idea would be to arrange a meeting with Mr Andrews as a fact-finding mission and with a follow up with any information that is required. Also, establish with Mr Andrews what Mr Phillips' role will be in the selection process and establish who else might be involved in the decision process.

❏ At the meeting with Mr Andrews, establish how previous contracts were secured, who the competitors are.

❏ Contact Mr Phillips in 6 months and pick the right moment to visit him and build relationships with all the people who may be involved in the consultant selection process.

If the target clients are not as forthcoming as the example above, then consider the following:

❏ The internet. Visit the company web page and find out from the director's report (often found on downloadable annual report) if there are plans for growth or change.

❑ If you are very much a regional player, then the local press will tell you what's happening in the local business. New recruits, especially at the high level, are often featured. This could generate a 'switch' from 'not buying' to 'buying' mode.

❑ If you are in a specific sector then get the appropriate trade magazine or paper. Track the movements of senior people; not just the incoming, but where the outgoing people are going?

❑ The local Chambers of Commerce may also have information valuable to you. You might, subject to funds, be able to commission them to do some research on your behalf.

❑ Carry out a survey. More of this is discussed in Stage 6.

The key point is to recognise that switch points occur, even in your existing and past clients. The important point is that you need to dedicate time to keep in touch with existing, past and potential new clients. You need to be able to see a potential switch point occurring. If you want to negotiate an opportunity to pitch for work, you need to gauge the right time to court the target clients.

For these clients, where you have identified a project, you may wish to create a client capture plan. This is no more than a report containing all the information obtained from your contact with the client or additional research. A good CRM package may be sufficient to capture all the information required.

The capture plan will assist you with valuable information, which may be of use later. Information to capture would be related to key people, their roles, business information and updates on any leads.

1.6   Creating a pipeline

A good business development programme will have a 'pipeline' or a record of leads and enquiries.

A lead (sometimes referred to as a prospect) is project specific. It is where an actual project is identified. It is not a client. Clients can generate many leads. A lead is the first stage in the pipeline. The client might not be aware of you at this stage, or if he is, he has not asked you to do anything.

An enquiry is again project specific when the client has asked you (or enquired) to produce a proposal, price, initial ideas etc.

With time, you will build up your pipeline and you will be able to work out its

❑ current value;

❑ conversion rate from lead to enquiry;

❑ conversion rate from enquiry to job;

❑ trends such as success rates within sectors and success rates dependant on strength of relationship.

Points to consider when inputting data are as follows:

Probability of leads being converted to jobs:

❑ If it's an OJEU notice, the probability starts off at only 2% unless you are known to the client and already know about the project.

❑ In respect of other potential clients who don't know you (and not an OJEU notice) the probability should not be more than, say, 10%.

❑ If you are known to the client, then maximum probability should still be, say, 20%.

Probability of an enquiry being converted to a job:

❑ If you are in a tender situation, and you know the number of people tendering, then you are able to arrive at a probability of success. If there are four tendering then your probability of success would be 25%, unless you believe you have a better or worse chance of winning than the others tendering.

❑ If nobody else is being asked for a proposal then it's purely the probability of the job going ahead. Start off low and raise the probability every time you review the project, and it seems to be progressing well. Or, lower the probability if the project is not progressing or is likely to be cancelled.

Take also into account relationship values when considering the probability of success in converting leads to enquiries and enquiries to projects. You might look at simple criteria such as

❑ hot (good relationship)

❑ warm (developing relationship)

❑ cold (little or no relationship).

Alternatively, use a one-to-ten classification as shown in **Table 1.5**.

Using an enquiry pipeline

Table 1.6 is an enquiry pipeline for our fictitious architects practice Wren & Barry Architects Ltd. The table lists all the current enquiries being processed by the firm and 21 opportunities are listed.

The very process of putting together a list makes the firm address the knowledge it has. Many firms do not commit this to paper. All too often, each partner or director carries the knowledge in their head and is not shared as business intelligence. By creating your own enquiry pipeline, you will have a better understanding of likely workload coming in and where to direct your efforts to convert to projects. An enquiry pipeline, as shown in **Table 1.6** can tell you many things.

Table 1.5   Potential client relationship classification

	Potential client relationship classification
1	Client does not know you
2	You have communicated to the potential client (sent a letter)
3	You have spoken to the potential client
4	You have met the potential client
5	You have met the potential client several times
6	You have presented to the potential client about your business
7	You have presented to the potential client about this project
8	You have been asked to submit a proposal, bid or tender
9	You are in dialogue with the potential client after the bid or proposal has been submitted
10	You have secured the project and this is an existing client

Table 1.6 Enquiry pipeline for Wren & Barry

Enquiry pipeline for Wren & Barray

No	Client	Project	Probability % (P)	Fee Value (F)	Probability x fee value	Comments
E01	Brown & Jones	New extension to office	70	78k	54.6k	Won
E02	Heathlands Dev	Industrial unit	25	40k	10k	
E03	NW Health Trust	Refurbish waiting area	80	6k	4.8k	
E04	Brown & Jones	Office layout	70	6k	4.2k	
E05	Midshire Leisure	New squash courts	25	24k	8k	
E06	Midshire Leisure	Refurbish leisure centre	30	80k	24k	Lost
E07	Drake Pearson	Master plan	60	6k	3.6k	Lost
E08	SE Health Trust	New outpatients	15	110k	16.5k	
E09	Barker & Conlan	New warehouse	90	80k	72k	Won
E10	Bright Goods	New shop fit out	75	18k	13.5k	
E11	Barker & Conlan	Gatehouse relocation	80	5k	4k	Won
E12	Barker & Conlan	Refurbish staff canteen	40	22k	8.8k	
E13	SE Health Trust	Reception area	15	18k	2.7k	
E14	Brown & Jones	New boardroom	80	5k	4k	Won
E15	QP Plant Hire	New warehouse	30	45k	13.5k	
E16	Allan & Parker	Office space planning	90	15k	13.5k	Won
E17	City Builders	D&B office	30	45k	13.5k	
E18	Midshire Leisure	Refurbish spa area	50	34k	17k	
E19	City Builders	D&B industrial units	30	30k	9k	
E20	Bright goods	Warehouse	80	45k	36k	
E21	Allan & Parker	Reception area	25	6k	1.5k	Lost
		Total	Average 51.9%	718k	334.7k	

From **Table 1.6** you can see the following:

❑ The total fee value predicted if all the jobs came in would be £718k. This seems a good pipeline for a practice that hopes to have an annual turnover of £750k. Or does it?

❑ The average probability of all the jobs being converted to job wins is 51.9%, with just over a 50 : 50 success rate.

❑ If you multiply each enquiry by its estimated probability of success and then add all these 'discounted' fee values, then the table gives us a reduced total value of £334.7k; perhaps not so healthy for a turnover target of £750k.

❑ The table also shows which jobs have already been decided. We have 5 wins out of 21, giving a current conversion rate of 23.8%. At the same time, we are reporting three losses that give a current loss rate of 14.3%.

❑ The average job value of the whole list is £34.2k and the average job value of the five jobs won is £36.6K.

❑ The value of fees lost is £92k, with an average fee per job lost being £30.6k.

❑ If we now take away from the list the job wins and losses, then we have a revised total value of remaining fees of £443k, a revised average probability of 43.5%, giving a total discounted fee of £192.7k.

So apart from being a set of statistics, let us now see how we can use these figures to give us a feel for the health of the pipeline as it currently stands. We can now work that out:

❑ If we need an annual turnover of £750k, with a current discounted total of £192.7k, we have a shortfall of £557.3k.

❑ Another way to look at it is that if we always want a pipeline that could give us a £750k fee, the discounted total value always needs to be around £3.15m, allowing for a conversion rate of 23.8%. Suddenly we have a wakeup call for Wren & Barry!

❑ If all the enquiries were to follow the same profile as these first 21 enquiries, then our enquiry list would need to have around 92 enquiries (this being £3.15m divided by the average job value of £34.2k).

Of course, these figures do not account for some jobs coming straight in without going through the lead and enquiry stage. With well-established firms that look after their clients and their clients having a constant stream of work, this process is slightly watered down. However, if you do know your clients well then you will probably know what jobs are coming up and these would start to appear on the lead lists.

The process we went through above for Wren & Barry was just a snapshot in time. These figures really do come into their own, if you do the same exercise every month for several years. Doing this over a long period helps as follows:

❑ You will be able to show trends.

❑ You may be able to predict that you need to have a certain amount already secured by a certain time each year to feel comfortable to end the year on target.

❑ This also helps the sales team to report to management. Senior management will have more comfort from a process-driven report than just a 'feel' about how it's going.

❑ Also I have found this kind of exercise useful for firms who are looking to sell or merge. It is a good way in 'evaluating' goodwill and likely future revenue.

Having introduced many firms to this process I have seen a dramatic change of focus and purpose in sales and marketing effort.

Many people get concerned about probability ratings – are they too optimistic or pessimistic, or how can they overcome different sales people's views on probability. The key point here is that each person should 'guess' as accurately as possible, given all the influences on success. Also, the important point is that they should be consistent. If they are naturally optimistic then that's fine; the statistics show averages and that is then balanced out.

The other point that the enquiry list tells you is the spread of your enquiry portfolio. Taking the Wren & Barry list as an example we can see the following:

❑ Bright & Jones account for three enquiries and for two of the five wins.

❑ Baker & Conlan account for a further three enquiries and also two of the five wins.

It might also be useful to put alongside each enquiry the client account person within the firm handling the enquiry, so you can see who is bringing in the work. I once worked for a firm where the senior partner always seemed to bring in the 'flagship' projects, but on closer inspection his partner was bringing in more fee income, be it through medium and small projects.

Also, it might be useful to look at market sectors, both from the point of view of the project and the client. This might confirm what you already know, that is, which sectors are featuring in your enquiry list and which are being more successful. This could be a reflection on how the market perceives your firm. Are you, for example, seen as a commercial sector firm and is that why you are not breaking into the leisure and health sectors?

Using a leads pipeline

In the same way as the enquiry pipeline is useful, so is the leads pipeline. The format for the lead pipeline would be the same as for the enquiry. The key points to consider when creating a lead pipeline are as follows:

❑ List all the opportunities that you become aware of, which you believe, at the time, fit your skill set and are worth pursuing.

❑ A lead is a project opportunity which you have identified but have not been asked by the client to do anything about. Once you have been asked by the client to provide a sketch, fee bid or some input, then it becomes an enquiry. A lead does not have to be for an existing client.

❏ The probability of success will be much less than for an enquiry. There is not only the probability of the project progressing to fruition but also the probability of you winning the work. Also for non-existing clients they might not know you and so will have a low relationship value rating.

To demonstrate the value of the lead enquiry list in conjunction with the enquiry pipeline, let us assume the following for our Wren & Barry architects example:

❏ There is a conversion rate of 10% from lead to enquiry status.

❏ Some enquiries come straight in as enquiries and never feature in the lead pipeline. Let's assume that 60% of all enquiries fall into this category.

❏ Wren & Barry need a £750k fee turnover.

Fig 1.6 shows the complete pipeline summary and shows that to feed the required turnover, 350 leads need to be generated, leading to 88 enquiries resulting in 22 jobs, with an average fee value of £34.09k.

Having gone through this process, it gives some focus on the effort required to generate the turnover required. Of course, the figures will regularly change and in time might reflect a different portfolio. There might be a shift to tender for more public works which might have less probability of success. But, on the other hand, the public works could be for higher value projects, therefore increasing the average job value. Alternatively, the business might become more consultancy driven instead of project driven so that smaller assignments, but perhaps more of them, are generated. Either way, the pipeline will give a rough idea on how many opportunities need to be identified and pursued to arrive at a given pipeline. I say a rough idea because this system can only give an indication and reveal trends in potential workload. Even so, it is a good way to monitor the health of the future workload potential.

The marketing and sales effort needs to be continuous because of the usual long lead in periods. Some firms tend to concentrate on 'selling' when there is a downturn of work. Perhaps the best time to pursue work is during the busy periods so that leads and enquiries

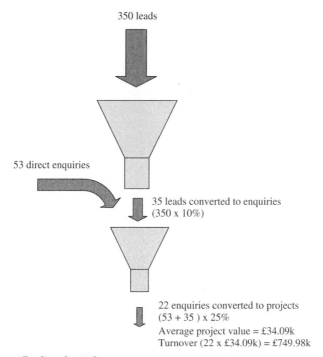

350 leads

53 direct enquiries

35 leads converted to enquiries
(350 x 10%)

22 enquiries converted to projects
(53 + 35) x 25%
Average project value = £34.09k
Turnover (22 x £34.09k) = £749.98k

Figure 1.6   Feeding the pipline.

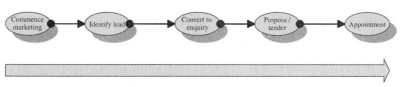

Commence marketing → Identify lead → Convert to enquiry → Proposa / tender → Appointment

Time delay from commencement of marketing activity to appointment. This could be weeks or years. Some opportunities may come into pipeline as enquires. especially from existing or past clients.

Figure 1.7   Time delay from marketing activity to appointment.

can be converted in time for a downturn in workload. **Fig 1.7** indicates the sequence of events resulting in different lead in times for each project. This emphasises the time lag between marketing and sales effort and the projects being identified and secured. The more leads and enquiries you can generate to feed the pipeline, the better your chance of converting a greater number to projects.

In large organisations, there may be many people with the ability and opportunity to generate leads and enquiries. In such situations, it is useful to have someone, perhaps a business development manager or director, tasked with helping others with their individual pipeline development.

Managing effort for pipeline development

In many organisations, there is never enough time to devote to creating enquiries to feed the pipeline. Many senior people within an organisation are well connected to current and past clients or other consultants who may have important market intelligence, but never have sufficient time to make the calls. If this is the case within your organisation, then you need to do the following:

❑ Put someone in charge in managing the process. This person should have reasonable internal authority to make things happen.

❑ Prepare a strategy for re-engaging with contacts that may be useful to you.

❑ Share out the tasks and make the tasks manageable. Start small and work up.

❑ Get a proper CRM system in place.

Irrespective of the systems you wish to employ, you need to manage each target client. **Table 1.7** shows the client relationship summary for one of Wren & Barry's clients. The table lists all the existing relationships and the quality of that relationship. There is also an action section for developing the relationship further.

Table 1.8 extracts some of the information from **Table 1.7** and gives an overall feel (value) for the client relationship. In this example, it is coming out as 'good'. Again there is an action for continuing the relationship building.

These tables, or similar reports on client relationship, should be reviewed regularly to make sure that relationships are not getting cold.

Table 1.7 Client relationship summary for a Wren & Barry client

Client relationship summary for a Wren & Barry client

Client: Barker & Conlan health products		Key client account manager: Brian Oliver				
Existing relationships		Status:	A: Instruction giver B: Influencer C: General	H: Hot (will always think of us) W: Warm (needs encouraging) C: Cold		
Name	Job title	Relationship Status	Relationship	Prime contact	Also known by	Comment
Richard Huggins	Managing director	A	H	Brian Oliver		
Mark Johnson	Finance director	B	W	Henry Farrow	Brian Oliver	
Gary Bleach	Facilities manager	B	W	Brian Oliver		Introduce to Jane
Ruth Wright	HR director	C	C	Brian Oliver		
Barry Brown	Operations manager	B	C	Brian Oliver	Henry Farrow	

Targeted relationships					
Name	Job title	Action	By who	By when	Update
Phillip Garner	Operations director	Invite to our next breakfast seminar	BO	Within 4 weeks	Accepted invite

Actions

Key actions/objectives over next 3 months	Event/opportunity to enhance relationship	
Check Phillip Garner attending breakfast seminar	Gary Bleach wants to visit office furniture showrooms	
Recent enquiries/referrals/opportunities created	Review date	Signed client account manager
Gary Bleach considering new office layout for accounts dept		

Table 1.8 Client relationship overview for a Wren & Barry client

Client relationship overview for a Wren & Barry client

Client: Barker & Conlan health products	Key client account manager: Brian Oliver

Number of existing relationships comprising of	5 of which			1 Hot	2 Warm			2 Cold		
	1 Instruction giver				3 Influencers			1 General		
	1 Hot	0 Warm	0 Cold		0 Hot	2 Warm	1 Cold	0 Hot	0 Warm	1 Cold
	1	2	3	4	5	6	7	8	9	10
Overall feel for relationship	Bad		Poor		Good X		Very Good		Excellent	

Possible relationship-building events to be considered (in addition to regular telephone/letter/e-mail contact)

Activity	Who	Action by	By when
Presentation to them			
One to one (lunch/dinner/drink)			
Group activity			
Invite to event			
Joint initiative with them			

Summary checklist

By the end of Stage 1 you will have examined your current client portfolio and found

❑ the client mix by client quality (gold, silver, or bronze);

❑ work intake by market sector;

❑ the spread of income through client portfolio and examined potential vulnerability.

You will have worked out a strategy with

❑ priority one being existing clients;

❑ priority two being past clients;

❑ priority three being new clients.

You will start selecting target clients based on the criteria of

❑ location;

❑ market sector;

❑ size of client.

You will look at the type of client you want to pursue to give a balanced client mix. Your selection process will take into account effort versus reward and take a look at potential lifetime income from clients. You may wish to pursue

❑ public sector;

❑ private sector.

When contacting potential clients you will be aware to pick the right moment to increase sales activity to particular target clients. You will become aware of the following:

❏ Switch points when the client might switch from not buying ('Off') mode to buying ('On') mode.

You will be conscious of the need to build relationships and to start recording

❏ a pipeline that includes leads and enquiries;

❏ probability of success based on relationship with target client and amount of competition.

You will also regularly review each existing and past client contact so that you can

❏ maintain current relationships;

❏ identify who needs to be contacted;

❏ have a campaign of activities and events to maintain contacts and continue to build relationships.

Stage 2: Identifying the needs of the target client

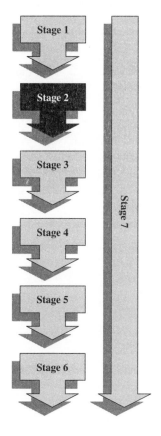

2.1 Why do you need to know the client's needs?

2.2 Appropriate person to contact

2.3 The first Meeting

2.4 Look for visual and auditory clues

2.5 Establishing the client's needs

2.6 Listening skills

2.7 Establish the stakeholders and decision makers

2.8 Expanding the range of contacts

2.9 Establish selection criteria

2.10 The next step

Stage 2: Identifying the needs of the target client

2.1   Why do you need to know the client's needs?

Many professionals think that they offer the service that the client needs. Then they wonder why they are not appointed to deliver that service. Every client is different and so they deserve, and often expect, a service tailored to suit their needs. If you do not know their needs, how do you expect to convince the client that you can deliver what they want?

The best way to find out the client needs is to ask the clients themselves. This is easier said than done. Let's assume that you have drawn up a list of potential clients that you wish to work for. Now you need a structured approach to attracting their interest and gaining an opportunity to meet with them. Let's assume that the target client has not heard of you and your organisation (see Stage 4 for promoting yourself) so you need to do the following:

❑ Find out the appropriate person to contact.

❑ Find a way to make contact with that person.

❑ Create an opportunity to facilitate an appointment.

❑ Meet with the target contact.

❑ Develop a rapport and elicit valuable information.

❑ Follow up communications.

Good client service

Good client service is whatever the client says it is. It is not measured against a set of criteria made up by the marketing department or the sales team. If you can identify the needs of the client and provide a service that satisfies those needs, then perhaps that is providing a good service.

However, the needs of the client will probably change because

❏ their expectations move on;

❏ the client interface moves on;

❏ repeat clients tend to become more sophisticated and demanding.

A group of customers may have the same needs but they want those needs satisfied in differing degrees. Therefore, it is not just good enough to establish what the needs of the client are; you also need to establish the relative importance of the needs.

2.2   Appropriate person to contact

Within an organisation there are generally four types of persons who will be useful to you (see **Fig 2.1**). These are as follows:

❏ **Level 1** – the person who signs your appointment and signs the cheque. This will be someone like the chairman, managing director, works director and facilities director.

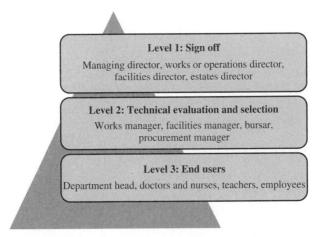

Level 1: Sign off
Managing director, works or operations director,
facilities director, estates director

Level 2: Technical evaluation and selection
Works manager, facilities manager, bursar,
procurement manager

Level 3: End users
Department head, doctors and nurses, teachers, employees

Figure 2.1   Three levels of possible contact within an organisation.

❑ **Level 2** – the technical person who has the responsibility to select the appropriate person for the project. This will be someone like the works manager, facilities manager, procurement officer, works engineer estates manager.

❑ **Level 3** – the end user who will use the facility once completed. This will be someone like department head, head doctor/nurse, headmaster.

❑ **The influencer** – there is also, within organisations, the influencer. This could be someone like the finance director or human resources director who may not have a direct involvement in the process but may have a degree of influence in the selection process.

You will find that unless the organisation is very small it will be very hard to contact, from a cold start, the level 1 person. Also the level 3 people will not have the appropriate influence and knowledge of the process to be very useful to you at the very beginning of a relationship. The persons to target are the level 2 people. Win them over and they will sing your praises and introduce you to the other two levels and influencers. You will probably need to win over all levels eventually.

The first task for the professional is to establish a relationship within the target client organisation. If you do not establish the roles of the various people within the organisation, you may run the risk of 'selling' to the wrong person. You need to find out quite early how the target client goes about deciding who they use. Therefore there is a need to establish the following:

❑ Who is the decision maker? Does that rest with level 2 who just gets level 1 to sign off? Or does level 2 make recommendations to level 1 who therefore also needs to be nurtured early on?

❑ Who will be the users, who are the most powerful?

❑ Who are the influencers?

The next stage is to find out who within the target organisation are the level 2 people. The best approach is to phone up and find out.

This can be delegated to someone in your organisation who

❑ has a good telephone manner;

❑ is able to strike up rapport;

❑ has a knack of finding out information without being a pest;

❑ is methodical and can accurately record the information obtained.

You or the person you have delegated to make the phone call should have a script or structure to fall back on, but don't make it sound as if you are reading. A typical telephone conversation might be as follows:

You: *Hello, I wonder if you can help me.*

Receptionist: *I will try to, how can I help?*

You: *I am trying to establish who in your company tends to appoint consultants like architects, engineers or project managers?*

Receptionist: *Oh I don't know, I am new here.*

You: *No problem. Do you have a procurement department?*

Receptionist: *Yes, do you want me to put you through?*

You: *Yes please, who will you be putting me through to?*

Receptionist: *I will try Mr Cuthbert as I know he is in.*

You: *Thank you.*

Mr Cuthbert: *Cuthbert here, procurement department, can I help you?*

You: *I hope so. I am trying to establish who in your department selects architects, engineers or project managers?*

> Mr Cuthbert: *We only handle the contracts here, you will probably need to speak to Henry Smith who deals with site services or John Fairbanks who is within the estates department and tends to deal with capital works and maintenance. Keith Phelps, who is the facilities manager, tends to get involved with the interior design and office layouts.*
>
> You: *That's very useful. Thank you.*

So from this conversation you have learnt and can record the following:

❑ John Fairbanks is probably the key level 2 contact.

❑ If selling interior design or space planning, then Keith Phelps might be the appropriate level 2 contact.

❑ If selling M & E services then Henry Smith might be the first point of call as a level 2.

❑ Mr Cuthbert is now an established contact within the procurement department and could be an influencer.

The main objective is to find out who is in the level 2 position within the target organisation responsible for appointing or selecting professional services organisations. If the operator is not sure, then ask to be put through to a secretary or PA or clerical assistant in a particular department that seems the most relevant, e.g., estates or procurement department.

In large organisations, you may find that the person you are looking to track down might be based on another site and there might be several people who might fit the role (for different regions or for different types of work). It is important to record everything you are told at this stage.

Remember that the process is all about building a relationship with your target. Don't rush to sell at this stage.

No names policy

Increasingly, businesses have a no names policy. This is where they will not give out any names and not pass you through unless you give them the name of the person you want to speak to. Usually they will tell you to write in to a job title like the estates manager. This is almost a waste of time because

❑ you are not able to follow up unless they respond to your letter;

❑ they are very unlikely to respond unless you are exceedingly lucky in gaining their attention at the right moment and you are offering them exactly what they need and that they need it now.

You are better off by doing the following:

❑ Searching the company website and finding out some names from there. The names in the company annual report might give you some high level (level 1) names. You could then try to be put through to their PA and ask if their boss is the right person to speak to (unlikely) or do they delegate the call to someone (likely), and just hope that they might give you a name or put you through.

❑ Alternatively, you could write in to the job title and invite them to a seminar, workshop or visit, which would be of interest to them. This might tempt them to make contact and establish communication.

❑ Another possible way of finding out names is if you come across other consultants who work for the company and ask them who they are dealing with.

The first communication with the target contact

If you were to try and telephone the target contact then it is very likely that you might get the following responses:

❑ Be told by their 'gatekeeper' (secretary, PA or assistant) who picks up the phone that the person does not take cold calls and you will need to write in.

❑ If you actually get through to the right person, they might well tell you to write in with details and they will come back to you if interested.

So to overcome this potential problem you need to write into your target contact. Then, when you call and you are asked by the gate-keeper what your call is about, or will the target know what you are calling about, you can truthfully say that you are in correspondence with the target and you are following up on some detail point. If you fail to get past the gatekeeper then you need to phone up very early or late or at lunch time when the gatekeeper will not be there. Senior people will often start early or work late or both and will probably answer the phone when the gatekeeper is not around.

The first written communication

The first communication needs to hold your target's interest. The following are possible examples:

❑ If you are trying to expand within a sector you are already working in then send the target a project sheet, case study or article about one of your recently completed projects. A potential client will want to know what his competitors or rivals are up to, but you do need to make it relevant.

❑ If you are transferring appropriate skills from one sector to another, then again do as above; but even more importantly show how the skills and experience you have obtained in a recent project are appropriate to a possible project your target might have.

❑ If it's none of the above, then it might be that you are able to offer something that you know your target might be interested in. Perhaps a service that is now in need because of a change in regulations or that may be able to save the target client money in the short term.

Whatever you write, assume that it will be just glanced at and filed away in the nearest bin. That does not matter as it will serve its purpose and hopefully get you through the gatekeepers and be a potential talking point with your target.

This first letter will have the following objective:

❑ Preparing the way for a follow-up telephone call where you will be looking to make an appointment.

❑ Not looking like a mailshot. So it needs to be very tailored and bespoke.

❑ Introducing a subject or current issue that will grab the target's attention. This could be related to a strategic activity which you have discovered from your earlier research.

❑ Briefly explaining how your firm has the appropriate experience and skills to help the target with the current issue.

❑ Suggesting a meeting and advising them that you will be following up with a telephone call to arrange for a meeting.

This first letter must not be longer than one page. Every paragraph should be relevant to the potential client and make them want to read more. Below is an example of a typical first letter.

Dear Mr Fairbanks,

Re: 15% energy savings achieved at recent office development for Brown & Jones

You may have read in the local press that Brown & Jones have just moved into their new offices. The facility was designed by us and we were able to achieve

❑ *a 20% reduction in floor space for the same number of employees;*

❑ *a cost saving of 15% on annual energy consumption;*

❑ *an enhanced working environment, endorsed by the employees as a good place to work.*

I attach a case study which explains how we were able to achieve these results for Brown & Jones and I am sure we could bring the same skills and experience to a new office environment, which I understand you might be considering in the near future.

I will therefore telephone you in the next few days to establish if you would welcome a meeting to explore the many ways Wren & Barry Architects could be of assistance to you on your business campus.

Yours sincerely

That first phone call

Remember that you are not trying to sell at this stage and that comes later in the process. You will want to start developing a relationship and to express a desire to work with your target clients.

When making these cold calls remember the following:

❑ Sound enthusiastic; be in the right frame of mind.

❑ Make out that you have done your research (and you should have!) by mentioning facts which are specific to the potential target client.

❑ Make it sound that you know his industry or sector and you are targeting him specifically. Make him feel special and important.

❑ Do not sound like you are reading from a script.

❑ Engage in a conversation, not a sales pitch.

❑ Listen to what they say and make relevant notes.

❑ Don't talk too fast, and be clear and to the point.

❑ Don't leave messages and don't ask them to phone back, keep the initiative to make the next call.

The professional must realise that the aim of the initial contact is to establish rapport, build a relationship, identify the right opportunities to pursue and establish the client needs. This may take years of keeping in touch and developing a relationship.

During that first phone call you will have a very short moment to try to establish a need to meet and try and obtain an appointment.

So prepare for all eventualities. Have a few notes written down. Even have a rough script prepared, but don't read it out word for word as it will be obvious to the target client.

There are two possible objectives for this first phone call:

Objective one: If you already know that the target has a property portfolio and is frequently in the market, then you will be looking to make an appointment. You will have written to the target and will be following up a few days later so the letter is still fresh in their minds (unless it was filtered out by the PA!!).

Objective two: If the target is an infrequent user of your type of services, and you are not sure if there is a current need, then you could use the phone call to establish the need for the appointment at this stage. You might take the view, if there is nothing on the horizon for a few years, not to meet but to keep in regular contact. In this case, try to find out as much as possible at this stage and enter the information into your database. Try to find out

❑ if there are any projects coming up;

❑ what triggers their start (new process, office move, new financial year, etc.);

❑ who will be responsible for selecting consultants and contractors?

Keep the first call short or you might get your target annoyed. Below is an example of the type of call you could be making:

Mr Barry: *Good morning Mr Fairbanks. It's just a quick call. I know you are a very busy person. Can you spare a few minutes?*

This tells the target client you know his name, it's only going to be a short call and is it convenient for him to take the call now.

> Mr Barry: *My name is John Barry from Wren & Barry Architects. You may recall I sent you a case study a few days ago about our project for Brown & Jones, their new office building. I understand you are the right person to be talking to regarding new projects, is that correct?*

This tells him that you have already sent something in the post so hopefully he won't cut you off by requesting you to write in with your details. Also it tells him that you have done your research and found out that he is the right person to talk to (of course if you now find out he is not then get the right name from him, thank him and say goodbye.)

> Mr Barry: *I am calling to establish if there are any projects of a similar nature, or indeed any new build or refurbishment work, we could perhaps help you with in the future?*

You will now get one of several responses:

❑ Yes, I do have a need and would welcome a meeting.

❑ No, we don't have a current need but might do in a year or so.

❑ No, we don't have a need and won't have.

❑ We have consultants we use and are very happy with and are not looking to change.

At least this call will have established the current situation. If there is no future need or they are happy with their current consultant's then move on to other targets.

Once you have obtained the information you need then thank the target and ask for permission to keep in touch. Agree on when you can call again. If the project is several years away, then every 6 or 9 months should be ok. Don't make it longer because the situation might change and indeed your contact might move on and you will need to start the process all over again. If the project is less than a year away and the contact does not want to meet at this stage try to agree to call again in say 2–3 months.

Always try to obtain some relevant information like the internal name for the project. By doing this you can pick up in a few months' time that much easier. For example,

> *Good morning Mr Briggs. You might recall we spoke 3 months ago about Project Riverside, the office relocation you were considering.*

This tells the target that you did call him and you already know about the project. You can now continue with

> *We agreed that I was to call you again in 2 months' time to check if there was any progress with the project.*

This reminds the target that he had agreed you could phone again, so you are not seen as a pest.

If he tells you that the project has been delayed you can say

> *Yes I remember you saying that the project might take some time to get started. Are there any other projects that you or your colleagues are dealing with that might happen earlier?*

It is important to keep up the interest and keep a dialogue going. With regular contact you will get to know when the project(s) might happen and hopefully build rapport with the target so that getting to meet with him at the appropriate time will be easier.

During this stage you will be trying to qualify and track leads. Use this time to gain more information from your target client about the potential lead. Try to do the following:

❑ Put a value to it.

❑ Try to scope the size and complexity of the project.

❑ Establish how the project fits into the whole scheme of things. Does something else need to happen before the lead becomes active and the client enters the buying mode?

❑ Get to know about the person you are talking to. Learn about his interests and views on the project.

During these catch up telephone calls, you will eventually be getting closer to the time when the potential client switches from 'not buying mode' to 'buying mode.' Also in the interim you will have, with their permission, posted or e-mailed case studies, updates on good practice within the industry sector or interesting information to keep your name in the frame.

So when you feel that the timing is right arrange for a meeting. If you have any choice in the matter make it mid-morning. This gives the potential client time to deal with any problems that arise first thing and does not encroach on his lunch break.

2.3   The first Meeting

Preparation

Before the meeting, consider the possible structure and prepare. You will need to do the following:

❑ Do some research on the business. If you did some during the selection process then go back on the internet and double check if there is any recent news. A good place to start is the recent press release section, which will give you an idea of what's happening and possibly arm you with some small talk in the opening part of the meeting.

> *I saw from your recent press release that you have a new works director. Has he settled in now? Has he introduced any new ideas?*

This will not only open the conversation but also allow you to obtain some more information during the meeting.

❑ Read the latest annual accounts which are usually on line and part of the company annual review. This will tell you how the company is doing, and might also tell you what investments they might be looking to make, which could require construction work.

❑ Call the marketing department of the company and ask for some brochures about the business and its products or services.

At this stage do not pre-determine the type of service you will be offering the target client.

Arriving

The main point is being on time!! I always think that the last 5 minutes of the journey seems to take 30 minutes. Allow sufficient time to find parking (always try to pre-book a visitors place if available).

I always try to arrive at the reception about 15 minutes early and tell the receptionist I am early and not to announce my arrival until 1 or 2 minutes before the agreed time. I use these few minutes to examine the reception. This includes the following:

❑ When signing in look at the signing in book and see who else is visiting.

❑ Often there is a file in the reception which has media cuttings. Take a few minutes to read.

❑ Look at pictures and certificates on the walls. This will tell you something about the company culture. Perhaps there are pictures of employees doing charitable deeds, certificates declaring that they are Investors in People or in The Times top 100 best companies to work for, etc.

While waiting, refrain from sitting in low soft seating which is so hard to get out of. Best is to stand and be ready to greet your potential client. Also if offered a drink politely decline at this stage unless you are being offered the drink for when you are going into the meeting. Once in the meeting always take up the offer of a drink. This makes it more social and eases you into the meeting.

When the time is right ask the receptionist to let your contact know you have arrived. Hand over your business card so that there is no mistake with your name and where you are from.

First impressions

We have all heard the phrase

You never get a second chance to make a good first impression.

So let's assume you have been collected in the reception by the PA and are following him to your potential client's office or meeting room. To prepare you will have

❑ used the facilities while waiting to tidy your hair and clothes, especially if you have been travelling some time or walking outside in strong winds;

❑ the appropriate file or notebook in your hand, so you will not have to waste time looking in your briefcase once in the meeting room;

❑ made sure your hands are dry and not clammy ready for the hand shake.

When shaking hands use a good firm grip. Do not overdo it with a strong, 'bone crushing' handshake. You might wish to try to use your other hand to briefly grip or tap the other persons shaking arm just above the elbow. This technique is often used by politicians and seems to add to the greeting process. If you use this technique then try to make it as smooth and natural a process as you can.

When you are introduced to the potential client you will need to have some initial small talk but very early on you will need to

❑ thank the person for seeing you;

❑ reconfirm the time allocated for the meeting;

❑ reconfirm the subject of the meeting.

You will then outline

❑ the structure of the meeting;

❑ clarify your objectives;

❑ emphasise to your potential client the benefits he will get from investing time in the discussion.

By doing this you will be indicating to your potential client that you are well structured, professional and will not waste his time.

Improving your chances of a good meeting

Your objective at the first meeting is to build the relationship so that future meetings are easier to arrange and the potential clients start to trust and give you worthwhile information.

There are five points to remember, which will help control the flow of the meeting and establish rapport. These are the following:

❑ **Don't talk too much**. The professional will be tempted to do all the talking. To get the most out of the meeting the talking should be evenly balanced. When you see the client waiting to speak, pause and let him into the conversation. Even have some questions prepared to bring him into the discussion.

❑ **Build empathy**. Try to put yourself in the client's shoes. What will he want to know? What questions will he want answered? Also use his type of language, use his industry jargon, and show him you know his business, its activities and what could be his issues.

❑ **Come over as a professional**. You are selling a professional service so you must convey the professionalism of your firm. Be professional in your approach and how you handle questions. Never criticise your competitors.

❑ **Be enthusiastic**. An enthusiastic person will make the meeting enjoyable; it will give the client more confidence in your ability to service his needs. Above all, enthusiasm is contagious. No matter how the meeting is developing never let go of your enthusiasm.

❑ **Prepare for the next meeting**. It will be very rare to get a commission at the first meeting. Therefore, as you go through the meeting prepare the way for follow-up meetings. Make notes of any actions that come out and make sure you follow up as soon as possible. You want to be seen as efficient and professional. As you are establishing the client's needs start thinking how you can follow up by perhaps visits to other projects, to your office or bringing a colleague along for a follow up meeting if there is something particular you can latch onto. Always leave the initiative with yourself, be proactive and never leave action with the client. As soon as you have left he may be inundated with problems and never get round to you again. During the meeting prepare the way forward for future meetings.

2.4   Look for visual and auditory clues

You can learn a lot from watching and listening to your potential client. Paying attention to what you see and hear can help you develop rapport, which would then lead to a higher possibility for a successful outcome in the business relationship.

Building rapport

It has been established that when people are like each other they tend to like each other. Rapport is the process of interaction, and good interaction builds good rapport.

Communication between two people meeting each other incorporates

❑ **7% words** and revolves around experiencing and associating content and key words;

❑ **38% tonality** which is all about our voice and relates to its tone (pitch), tempo (speed), timbre (quality) and volume;

❑ **55% physiology** which is to do with our body language, this being our posture, gestures, facial expressions, including blinking and breathing.

It is suggested that rapport is enhanced if we mirror the communication qualities of the person we are talking to. This should be carried out in a subtle way and not made so obvious that the other person knows what we are doing.

Practitioners of neuro-linguistic programming (NLP) have developed this to a fine art. The professional service provider, who wants to hone his skills, would be well advised to take a course in NLP. NLP suggests that to build rapport you should try the following to mirror and match your potential clients:

❑ Posture

❑ Facial expression and blinking

❑ Breathing.

Obviously you will need to do this gradually and not make it obvious.

Use of language

Everyone is different and how they evaluate the world around them is made up of many clues and references. These can be basically broken down into the following:

Visual (seeing) people who usually tend to:

❑ sit forward in their seat;

❑ are well groomed, neat, organised and orderly;

❑ memorise by seeing images;

❑ not distracted by noise;

❑ have difficulty remembering verbal instructions because their minds start to wander;

❑ are concerned with how things look.

Auditory (hearing) people usually tend to:

❑ be distracted by noise;

❑ respond to tones of voice, or set of words;

❑ be told about proposals rather than shown;

❑ memorise by procedures and sequences.

Kinaesthetic (feeling) people usually tend to:

❑ move and talk very slowly;

❑ memorise by imaging 'doing' something;

❑ stand closer to people than the visual person;

❑ like things that 'feel right'.

Most people will rely on all these clues or references and most people tend to be dominated in one of these areas.

By listening and watching carefully, you will be able to establish which area your potential client tends to concentrate on. Once you have established this you can use the appropriate language to suit your potential client's outlook on the world.

You could use phrases such as the following:

For visual

❑ *You will like the look of this.*

❑ *Can you imagine what it will be like?*

❑ *This solution is crystal clear.*

For auditory

❑ *You will like the sound of that.*

❑ *That view rings a bell with me.*

❑ *I am sure the solution will harmonise with your requirements.*

For kinaesthetic

❑ *You seem to be grasping the idea.*

❑ *You can tap into the other ideas.*

❑ *It seems to feel right.*

Table 2.1 shows words that you could interweave into your conversation depending on your potential client's predominant trait.

The English language is one of the most extensive in the world with more than 300 000 words. In normal speech we tend to use only 1% of those available to us. If we listen to our potential clients, we

Table 2.1   **Example of predicates that could be used when speaking to potential clients**

Sample of words to use depending on potential clients' predominant trait		
Visual	**Auditory**	**Kinaesthetic**
Clear	Listen	Touch
See	Tone	Feel
Look	Say	Grasp
View	Near	Pressure
Illustrate	Talk	Hard
Picture	Sound	Solid
Watch	Tune	Unfeeling
Focused	Resonate	Concrete
Bright	Ask	Hit
Appear	Speak	Rub

may pick up on their use of in-house or industry jargon and we might also pick up on a selection of their favourite words or phrases. If you are able to pick up on these and use them back with the potential client it helps to build rapport because they realise that you are truly listening to what they are saying.

Is your potential client telling the truth?

Sometimes you are in discussion with your potential client and you are not sure if they are telling you the truth. This might be for sound business reasons. For example, you might ask

> *Are we the only ones providing you with a proposal for this assignment?*

The client might not be asking anyone else but does not want to give you the idea that the price does not need to be competitive and so he might say he is asking others to quote.

Research has shown that eye movement could indicate if someone is telling the truth or not. It's not a guaranteed method but one which you may wish to develop as a technique. What you need to do is the following:

❑ Ask your potential client some questions that requires them to recall things, and watch how their eyes move. Most people will move their eyes up to the left or right or sideways or even downwards when they are recalling information.

❑ Once you have established a regular eye response ask the 'difficult' question. If the eyes go to the same place then they are likely to be telling the truth. If, however, they respond differently then the probability is that they are not being honest with you.

2.5   Establishing the client's needs

The prime objective of the first meeting is to establish and understand what the potential client's needs are. Once you have been able to establish this you can ascertain if your service can satisfy those needs, and then move on to explain to the client the unique characteristics of your service.

This objective may be obvious and simple but it seems that too often the salesperson gets the process back to front. They are so keen to tell their potential client all about their services and the projects they have worked on, in the hope that this will impress the client and they will appoint them there and then. These salespeople tend to major on their service and experience rather than first establishing the customer needs. The salesperson who doesn't try to establish the needs of the client before selling his services runs the following risks:

❑ Irritating the client.

❑ Wasting the clients' valuable time, especially in the early stages of the discussion when they may be willing to give information about themselves and the needs and criteria of their business.

❑ Closing the door on an opportunity to sell to the client what the client wants. The client wants to feel important and wants to go through a pleasant experience where he is the focus of the meeting.

Specific needs

Your client will have a variety of requirements that need to be uncovered and understood. They will be a mix of the following:

❑ **Personal needs:** They want to be comfortable working with you and you will need to establish trust. They do not want, and indeed may have, a fear of the unknown and making mistakes.

❑ **Career needs:** They will usually want some job satisfaction and progression with their career or position within their organisation. They want a good outcome for the project.

❑ **Specific project needs:** This will be a whole range of requirements. But you will need to find out what the important ones are and ones that cause most concern.

During the conversation you will elicit the client's needs. This will be through questioning and listening.

Leading into the questioning

Although one of the objectives of the first meeting is to find out the client's needs it is always worthwhile to briefly outline to the potential client who you and your firm are. This brief introduction should only last for a few minutes. You will need to take your cue from the client. At this moment you don't know what the client's needs are, so refrain from doing a sales pitch. Don't, at this stage, take out brochures or power point presentations.

At this point tell the client you have researched his company and would it be alright to ask a few questions to clarify a few points. Once you have their approval try to ask some open questions. This allows the target to tell you what is important to them.

Try not to overdo the questioning, especially if you begin to notice the target getting a bit concerned about doing all the talking when they thought you had come to tell them how you were going to help them.

From this first meeting try to establish the following:

❑ What is important to the client organisation when embarking on a project? This could be cost and time certainty, getting stakeholders on board, or just making sure there are no surprises.

❑ What is the client process for getting the project sanctioned and possible time scales? Is there a budget or known size for the project?

❑ What experience do they have in the selection process of consultants and contractors?

❑ Do they have any retained advisors?

❑ What is the selection process for consultants or contractors?

Use your questions to demonstrate your capability

You can use the question session to demonstrate your capability and knowledge of the sector and target. Rather than saying, 'Why are you relocating?' you can say

> *Many of our commercial clients, who are deciding to relocate, are using the occasion to review their working methods and office environment. I would be interested to know what your key issues are in this respect.*

Have questions prepared

Have a series of questions prepared so that you don't miss out on key issues. However, don't be so keen to ask the next question that you don't listen to the previous answer and miss developing a key point revealed in an answer.

Use questions to coax out the information you want, but don't ask leading questions because you want to know the views of your potential client and not they endorsing yours.

A good well thought out question will help elicit the information you want. You may find it useful to do the following:

❏ Outline the nature of the questions especially if they are linked or on a common theme.

❏ Each question should be used to obtain information or an opinion. Both will be useful to you.

❏ Try to make your questions specific to your potential client. This will obviously show that you have researched the client's organisation prior to the meeting, for example,

> *I was reading on your website that the new technology you will be introducing will increase automation. How will this impact on your office space?*

❏ Only cover one topic per question. Make them simple to understand.

❑ Use something in the previous answer to lead into the next question. This again demonstrates that you are listening and understanding.

❑ Use open questions. That way you may get answers to questions you were going to ask and information you would not have expected to receive. You may only have a limited opportunity to ask questions so the open question is very useful. It also helps to develop a follow up conversation.

Harvesting information through questioning techniques

A good method of gathering information is by using your questions to move from the general to the specific. This process goes through a process of asking the following:

❑ **Open general questions:** This is a question that will give the client an opportunity to give you a full answer. This may also uncover information you did not expect and lead you down another avenue of questioning.

> Example: *I understand from our telephone conversation that you will be embarking on a major site re-organisation. What will be your first project?*
>
> Reply: *We will be looking at a new warehouse which will also incorporate a small amount of offices. This will be next to the existing external goods yard.*

Rudyard Kipling, in the following poem, identified how to commence an open question.

> *I have six trusty serving men*
> *They taught me all I know*
> *Their names are **What** and **Why** and*
> ***When** and **How** and **Where** and **Who**.*

An open or closed question is often determined by how you start it. **Table 2.2** lists typical prefixes for open and closed questions.

Table 2.2   Common prefixes to open and closed
questions

How you prefix your question will determine if it is an open or closed question	
Open question	Closed question
What	May
When	Do
Why	Can
How	Shall
Where	Are
Who	Is
	Would
	Could
	Have
	Will

❏ **Searching for follow-up questions:** This is where you have picked
   up something from the answer to the open general question. These
   could be further open questions.

> Example: *That sounds interesting. Who in your organisation will
>   be involved in the project?*
>
> Reply: *Well I will be project managing it internally. Our Managing
>   Director will give the go ahead once the budget and scheme has
>   been approved by the finance director and operations director.*

❏ **Closed questions:** Follow up with some closed questions to clarify
   the information you have been given and to check on some facts
   that have been divulged.

> Example: *Have you finalised a budget yet?*
>
> Reply: *No.*

❑ **Summarise:** Take time to summarise when you reach the end of a line of questioning. This has the benefit of showing you have listened, gives the client an opportunity to correct you or add extra detail and signals the end of that line of questioning and enables you to start a new topic.

> Example: *That's fine. So the warehouse is the first project and you will need to establish a budget and scheme before a go ahead to develop the project can be given.*
>
> Reply: *That's right.*

To illustrate the above method, **Fig 2.2** shows a possible sequence of questioning.

Open questions tend to stimulate conversation while closed questions tend to close the conversation down. Therefore having a mix of open and closed questions, as demonstrated above, will assist in the information gathering.

Depending on the type of information you are seeking, you may wish to consider giving your open question a numerical qualification such as

> *I understand that you have used professional service firms before. What would be the three key things you would look for in selecting a professional services provider for your next project?*

When asking questions be careful of the following:

❑ Asking a question which incorporates several questions. This may irritate the client and sometimes can be ambiguous.

❑ Asking questions in an interrogative, rapid fire or staccato way. Try to be conversational and if information crops up in the client's answer follow up that theme rather than sticking rigidly to a pre-determined list of questions in a set sequence

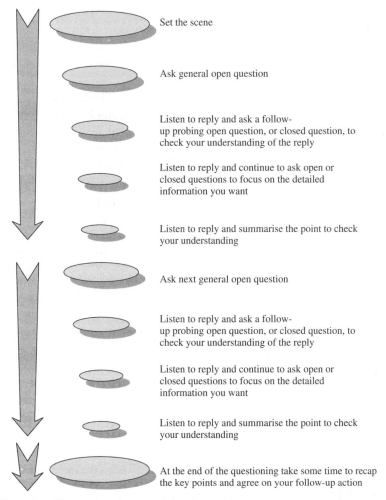

Set the scene

Ask general open question

Listen to reply and ask a follow-up probing open question, or closed question, to check your understanding of the reply

Listen to reply and continue to ask open or closed questions to focus on the detailed information you want

Listen to reply and summarise the point to check your understanding

Ask next general open question

Listen to reply and ask a follow-up probing open question, or closed question, to check your understanding of the reply

Listen to reply and continue to ask open or closed questions to focus on the detailed information you want

Listen to reply and summarise the point to check your understanding

At the end of the questioning take some time to recap the key points and agree on your follow-up action

Figure 2.2   Use a series of open and closed questions.

❑ Asking leading questions. A leading question is where a particular answer is suggested.

Example: *Why did you select* Industrial Design Ltd *for your last project? Was it down to price?*

By asking a leading question you may influence the clients reply, and therefore feedback might not be accurate.

Be careful not to do too much talking, you are there to find out information which will help you later to make a tailor-made presentation to the potential client. Many people talk too much because they are nervous. Alternatively they are thinking about the next thing to say and miss what is being said. Listening is the key to finding out what the needs of your potential client are.

Many professionals tend to give out too much information at the beginning of the discussion. They love to talk about their services or credentials because it is easy. You will obviously need to generate a discussion, which is far better than a question and answer session, as a discussion builds rapport and might reveal information that is important. The discussion should be structured such that there are open questions leading the discussion in the right direction to uncover the information you want.

Resist talking too much. Resist making a presentation because you don't know what the client needs are. You may do a quick overview about your firm to set the scene. You may respond to the client's questions during the conversation by quoting examples of work or showing a photograph of a project, which reinforces a point or desire that the client has raised.

Research has shown that beyond 2 minutes of talking about you the client is likely to lose interest. The more the professional talks, the more likely they will say something which is inconsistent with the client's needs.

2.6   Listening skills

You will need good listening skills to maximise your first meeting with your potential client. There are several ways you can improve your understanding of what is being said:

❑ **Make notes.** Making notes is a great way of improving your listening. Writing down key chunks of information boosts the entire listening process. You will need to refer back to your notes, so write clearly and don't use abbreviations you won't understand at a future date. It is always good to write up the meeting from your notes to create a file note. Also by taking notes you are signalling to your potential client that you think that what they are saying is important. However, do not take too many notes during the meeting because you may break up any rapport that may be developing.

You may find structuring your note taking will be useful. One method is to have three columns in your notebook. On the left hand side list all your prepared open questions. On the right side have two further columns. In one column you note down the reply and in the next column extract and write down any specific needs (see **Table 2.3**).

❑ **Clear your mind.** Don't try to engage in a meaningful conversation when you have other things on your mind. Try to clear your mind so that you can totally focus on the meeting. The last thing you want is for the potential client to see your eyes glazed over and giving the impression that you are not interested in what is being said.

❑ **Don't rush to speak.** Many people will interrupt and not let the other person finish. By doing this you may be deflecting the person from his line of thought and miss out on some important insight. Let the person finish and then pause before answering. This way you are also indicating that you have given some thought to what has been said.

When listening you need to indicate to the potential client that you are really interested in the meeting. You can do this by your body language. Remember to

❑ sit up straight in the chair and lean forward;

❑ make eye contact;

Table 2.3   **Possible layout within notebook for structured note taking**

Structured note taking		
Open questions	**Reply**	**Identified client needs**
'What will be your first project during your proposed site re-organisation?'	'The first project will be a warehouse with some office accommodation. This will be located next to the existing goods yard.'	Warehouse with integrated offices

❑ face your potential client full on;

❑ uncross your legs and arms.

While listening you can indicate to your potential client that you are actively listening by a nod of the head in agreement or just saying 'good point' or 'I agree'.

Your potential clients may not understand the design or construction process. They might not be as articulate as you. If this is the case do not show them up. Always show respect and show that you value what they say and are grateful for their time in explaining their needs to you.

Take time to also observe what is not said. Try to see the unsaid messages, read between the lines. You may then feed back into the questioning or conversation points what you have picked up on. Also it is a good idea to repeat to the potential client points that they have made. This reinforces the point that you are listening and paying attention. Also it acts as a double check to make sure you have properly understood the point.

When the potential client becomes aware that you are really listening they tend to feel rapport developing as the first step to the building of trust. Good listeners will develop their questioning techniques so that a conversation builds up rather than a string of short questions and answers. Once a good conversation develops, your potential client may open up and tell you important information, which you can use to prepare a proposal, making your offering more bespoke.

When listening you need to filter out all the signals. What is being said, how it is being said, the language used and body language. You will need to listen and categorise the information for use later when you shape your service to suit the client's needs.

When listening try to pick up on the following:

❑ **Approach.** The client might have views on how the assignment should be tackled. He might want a series of commissions to limit his exposure at any one time. Therefore he might welcome one stage at a time commitment. He might tell you how your competitors have tackled assignments in the past. Listen to the good and bad points about your competitor's previous work. Never criticise your competitors.

❑ **Personal emotions.** You should be able to pick up on the client's own emotions – his fears, his ambitions, his relationships with other stakeholders, decision makers and influencers.

❑ **New information.** You have carried out your research before meeting your potential client. Listen out for facts, figures, statistics and industry trends that are important to him and perhaps new to you. Feed back this information within your proposals at a future date to show you have listened and that these bits of information are important to you as well. Remember people buy people in the professional services world, so the more like your client you appear, the more comfortable he will be in selecting you.

Poor response

Some potential clients will not be happy with answering your questions. They may have said 'yes' to meet you to find out more about you and what your other clients (possibly their competitors) are up to. So be prepared to cut short the questioning if you detect that the potential client is getting irritated. Perhaps move on to describing your past projects and slipping in questions along the way such as

> *This project, for a manufacturer within your sector, found that he was able to reduce his warehousing requirements by managing his supply chain and delivery times to suit manufacture. As a consequence, we demolished his onsite warehouse and were able to expand his showroom and customer care centre. Have you found your warehousing needs reducing in recent years?*

Some clients will not want to divulge too much, especially at a first meeting. Perhaps company policy prevents the disclosure of such information. It is important to pick up on this and not to pursue a line of enquiry which is not going anywhere.

Sometimes the potential client will not know the answer and will not want to reveal that. Some people are just poor in answering questions and you may need to come back to a point later in a rephrased manner.

2.7 Establish the stakeholders and decision makers

Later on, when you are pitching for work with the potential client you will need to know who the key decision makers, influencers and stakeholders are. Take the opportunity at the first meeting to ask questions like

> *Who, apart from yourself, will be key within the selection process?*
>
> *Are there any key stakeholders who have strong views on the project and how it proceeds?*
>
> *What aspects of the project will they be most concerned about?*

If you can, and your potential client is being helpful, try to create an organisation chart, which shows who has the authority to sign off, who is part of the selection process and who will be the major influencers. Later, you will need to establish a way to meet these various people to promote your company and skills and services, and you will need to understand each person's view of potential benefits, e.g., finance director will be driven by saving money, management may be interested in improved efficiency and the users in the ease of operation.

You can also, at the first meeting, try to establish, or look for clues, to confirm that your contact has the authority to select or appoint. Or indeed establish how firm any likely project is.

Look out for the following:

❑ Reluctance to give any specific information; just looking for your ideas and opinions.

❑ Unable to give any kind of commitment or detail about the selection process.

❑ They will often lead the discussion and steer the conversation to the information they want.

❑ At the end of the meeting they will say things like

> *Thank you for the meeting. I will now need to discuss this with our . . .*

The possible reasons why non-decision makers, or those with no intention to change their current service providers, agree to meet (or possibly proactively call you to meet) are as follows:

❑ They want to challenge their current service providers, or check that they are still getting value for money from them.

❑ Use the meeting to gather information to make them appear more knowledgeable if questioned on the services by the senior management.

2.8   Expanding the range of contacts

You may find at your first meeting that the person

❑ is not the right person as he has no authority in the selection process;

❑ there are other people who are important in the appointment process;

❑ the client organisation is split into many divisions with different people in charge of selection for their own division;

❑ the person you are currently dealing with is about to retire, be promoted or is leaving the business.

Therefore, it is important to have a campaign to meet as many of the decision making people as possible. This needs to be done carefully so you don't undermine or damage your initial contact relationship.

Knowledge about your potential client is key to your success. All the information related to the buying process within the client's business is important to record, preferably on your database.

2.9   Establish selection criteria

During this first meeting you need also to establish the procurement process and any selection criteria that may be applicable. There is no point in preparing a pitch to commence a project in a few months' time if the client has an approved contractors list, or framework agreement or just been through a selection process for a supplier for the next 5 years or so.

Therefore establish the current position and what is required to get onto any approved list. It might be as easy as submitting company accounts and answering a few basic questions. It is better to know at the outset than find out later when you have invested lots of time and effort.

If the client does have a selection procedure try to find out what the criteria is and what the key issues are.

2.10   The next step

You need to come away from the meeting with an agreed action plan with a time scale.

It would be useful to maintain dialogue so that there are many options depending on what was discovered in the meeting. These could be as follows:

❑ If the meeting became a briefing meeting there will be a need to create a proposal. Don't rush to get to this stage especially without a well thought out brief. If the target was making the brief up as he was going along, there is a risk that it will not be well thought through and has not been agreed by others.

❑ In this situation try to arrange for another session, which could be a briefing meeting. This would also allow you to send through some questions so that the target could consult with others before meeting you again.

❑ If the project is not so imminent there would be time to arrange a presentation, which would be specific to the kind of project they had in mind. This would also allow you some time to thoroughly prepare.

The worst thing you could do is leave the meeting and not arrange some kind of follow-up action. Unless you decide there is nothing you could offer, or the target is well served by existing professionals who, he tells you, have served him very well.

Summary checklist

By the end of stage 2 you will have established the appropriate people to contact within your target client's organisation. You will also have established the following:

❑ The three levels of contact within the target organisation

❑ Who the key decision maker is

❑ Who the key users and influencers are

You will have established contact with the potential client and probably

❑ initially telephoned the contact and followed up with a written communication;

❑ having established the timing to be ideal, you will have arranged the first meeting.

In preparation for the first meeting you will have

❑ undertaken sufficient research into the client's organisation;

❑ prepared a series of open questions;

❑ prepared a quick overview of your business by way of introduction;

❑ prepared an outline for the purpose of the meeting.

At the first meeting you will

❑ look out for visual, auditory and kinaesthetic clues;

❑ listen carefully to what the potential client says;

❑ make sufficient notes to assist in establishing the potential client's needs;

❑ try to establish any selection criteria that may be in place;

❑ if time permits, try to expand the range of contacts within the client's organisation.

Stage 3: Shaping your service to suit the needs of the target clients

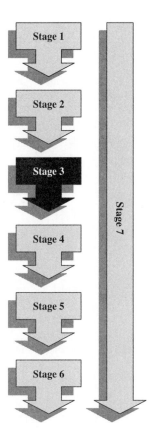

3.1 Review your clients' needs

3.2 Putting a mirror up to your client

3.3 Enhancing customer value

3.4 Features and benefits

3.5 Building trust

3.6 Differentiating

3.7 Consider your strategy

Stage 3: Shaping your service to suit the needs of the target clients

3.1   Review your clients' needs

After the initial meeting with the target client, there is a need to establish the following:

❑ Do we still want to work for this client? (See **Table 3.1** for an example of an analysis of clients' wants and needs.)

❑ Do we have all the skills and experience to undertake the project?

❑ Do we have a sufficient chance of securing the work? (If there are incumbents who are doing a good job then there are probably other clients where winning work would be easier.)

If the answer is 'yes' to all the above there is now a need to address the following:

❑ What additional skills or processes do we need to put in place to be able to convince the target that we can do their project?

❑ Is there anything that differentiates us from our competitors? Or indeed what do our competitors have that we haven't and puts us at a disadvantage?

❑ What do we need to work on to be able to convey to our target that we are the right people for them?

Quite often there is little need to drastically reshape how your business is run. But there might be some considerable work to do to convince the target client that you have no shortcomings.

Table 3.1   Analysis of clients' wants and needs

Analysis of clients' wants and needs		
Client: Mathews media	**Project: Office refurbishment**	
Key client wants and needs	**Our ability to meet wants and needs**	**Rating**
1 Designers to have experience in hi-tech offices	Have good experience and case studies. Can organise visits	Good
2 Need to produce high-quality visuals	Have in-house skills	Excellent
3 Need to get on board stakeholders. Departments have conflicting views on cellular offices and open plan	Experience in stakeholder involvement. Can show good examples of both	Good
4 Want to be in within 6 months	Resource problem, interior design department have little capacity for quick start	Poor
5 Have tight budget and want detailed cost control	Jones & Phelps QS available and experienced in this work, we have worked together on four similar projects	Excellent
6 Need price competition on construction	Local market is competitive for this kind of work	Good

Overview comments

❑ Can meet all the technical requirements of project
❑ We have a resource problem. Need to double check if completion within 6 months is critical for the client or nice to have. If critical then need to check on resource available. Unlikely at the moment to be able to free up resource for another month
❑ Other consultants are available with sufficient resources
❑ Who are our likely competitors?
❑ What is our probability of success?

Action

Check with client if 6 months is critical. If it is, we are not able to meet client's needs on this occasion

3.2   Putting a mirror up to your client

To know what your client wants to buy you must try to see yourself, and what you have to offer, through your client's eyes.

In most cases people will buy from similar people or organisations. They want to be able to trust the person and know that

❏ they will deliver what they promise to deliver;

❏ they have the resources to deliver;

❏ when there are problems they will come to the client's aid and help sort out the problems quickly and with least embarrassment and cost.

Therefore you will need to display, and in turn communicate, the attributes of your business that most match the attributes of your target organisation. If you are courting a charity or organisation that has strong social considerations (e.g., housing association), then you will need to demonstrate that you have consideration for social needs and the needs of those not so well off to have their own homes. You will need to emphasise your caring side. Conversely, if you were courting a hard-nosed industrialist who is struggling to compete with foreign competition then the points to emphasise are your ability to complete a project on or below budget, bring added value to help the client streamline and be more efficient to help compete in the highly competitive market.

Don't rush this stage

Once the professional service provider has met with the client and has had initial talks about their needs there is often a rush to get back to the client with some form of feedback. Sometimes clients don't fully know what their needs are until they have had several discussions with the professional.

It might be that you have met your level 2 contact and you still need to meet with the user groups to really understand the difficulties that need to be resolved. In some way, consider your time spent delving into your client's business as one of the following:

❏ A good foundation from which you can check if you can deliver all the needs of the client.

❏ Part of the relationship building process. The more time you spend with the client during these stages the better your understanding of their needs and their understanding of your approach.

❏ Learning about a business may also put you in a favourable position with other clients in a similar sector or market.

Unless the client has a definite project brief and is ready to consider your services, in which case you may go straight to Stage 5 (proposals and tenders), your aim after this stage is to go back to the client with evidence that you have all the skills and services he needs for his project (or part of a range of services if you are in a single discipline firm such as an architect).

An analysis of the situation

It may be useful at this stage to try to put down all the requirements the client needs from his professional team and analyse each part to establish your strengths and weaknesses.

If we continue the example from Stage 2 where we established that the client was considering an office relocation, we can list all the possible attributes that might be to our advantage when the client is evaluating our suitability. **Table 3.2** shows a typical strengths, weaknesses, opportunities, and threats (SWOT) analysis for the enquiry and highlights the key issues, which will need to be addressed in the proposal.

After the analysis you will need to address the potential weaknesses. In the example, we do not have all the required skills but can include others we have worked with to fill the gaps.

Now we can see a service delivery being developed, which would attract the client. Also if we have identified and resolved issues that the client was not aware of (e.g., the need for all the stakeholder involvement and signing up to a proposal early on), then this puts us in a strong position.

It is always worthwhile discussing the client's needs with colleagues and, where necessary, with other professionals whom you

Table 3.2   Enquiry SWOT analysis for a Wren & Barry enquiry

Enquiry SWOT analysis	
Client: CQS Investments Ltd	**Project Riverside: office relocation**
Strengths	**Weaknesses**
❏ Our experience ❏ Good client endorsements ❏ Our best office architect and his team will be available. Timing is perfect for our resource availability ❏ Good interior design department ❏ Good furniture audit team who have met the client to show how they can save space by good furniture selection ❏ We are local to the new site and have worked on the business park before ❏ We have good relations with the local planning department ❏ Our main competitors are not local to the site ❏ Our hourly rates would be less than those of Barrow, Jones & Bond ❏ We have other external consultants on board to form a multi-professional team. We have all worked together on many office projects	❏ Project is much bigger than our other office projects ❏ Our office does not look as high tech as some of our competitors, need to make changes before client visit ❏ Our relationship with the client, although good, has only been developed over the past 6 months. Not as good as Barrow, Jones & Bond
Opportunities	**Threats**
❏ If client goes the Design & Build way our client City builders would like us to join their team ❏ Client has agreed to visit our last project and then come back to tour the office ❏ Client expressed interest in giving us the feasibility study, as we are local to the new site and was impressed in our other feasibility studies for other clients ❏ Client has also hinted it was perhaps time for a change. Barrow, Jones & Bond seem to have undertaken most of their recent projects. Perhaps Barrow, Jones & Bond are becoming complacent	❏ Large national firms have expressed interest ❏ Barrow, Jones & Bond did their last office project ❏ Client expressed he might go the Design & Build route ❏ If project is delayed then our team might not all be available

regularly work with. Flesh out more strengths and weaknesses and start building up a case for why you should be chosen as opposed to others.

When putting together the case for why you are the best choice, you need to focus on the key issues rather than tackle a whole long list. The client needs to be convinced that you have been able to focus on the key issues, from his point of view, and have all the other minor points covered.

Has the client missed something?

When looking at your service delivery you need to be sure of the following:

❑ The potential client has told you everything. Perhaps the client is not aware of all the problems. That is why you need to speak, where possible, to the users (level 3).

❑ Your competitors don't have a winning edge, which you have not identified and therefore not even addressed.

❑ You use your professional skills to address aspects of the potential assignment that has not been identified by the client.

3.3 Enhancing customer value

Different clients place different emphasis on the various stages of service delivery. During the initial discussions with clients, be they existing, past or potential, you will be teasing out from them information to assist in improving service delivery. Don't assume that just because an existing client relationship is working well it couldn't be better. You must not be complacent. If you take your client relationship for granted then you might just let a competitor in when your client might be questioning the current relationship.

Your clients will experience many aspects of your business, from the sales pitch, various stages of service delivery, and the accounts department invoicing fees to the follow-up feedback sessions. Every time your client 'touches' your business it must be a good experience. Therefore the enhancement of client value should not just be seen as the delivery of the professional service but also as the back office team.

Client value is therefore the perceived benefits that the clients believe they receive from the service you will be providing. Client value can be evaluated by the difference you will be making on the client's business. Will you be helping the client in the following?

❑ To perform better with improved processes such as a new production line

❑ To become more efficient due to a better working environment, perhaps with better office layouts with better employee integration

❑ To lower running costs that could be passed onto the client's customers, perhaps by reducing overheads by reducing energy consumption

The bigger the impact, and across more areas, then the better the perceived value of the professional services.

To enhance the client relationship the professional services provider needs to define the precise nature of the added values to the clients. This could be on a client-by-client basis as well as market segment by market segment. Clients will often feel more comfortable in selecting a professional service provider who is very experienced in their sector because it is presumed that the issues are already known and there is no need for a 'voyage of discovery' where the service provider has to be educated at the client's expense.

Customer groupings

Through experience and client research it may become apparent that groups of customers will share common values and needs. This allows the service provider to sell a 'bespoke' service to some potential clients that were not prepared to have an initial exploratory meeting. Obviously, the identification of clients' needs is an important stage but sometimes the clients we are targeting are not prepared to invest the time for this exercise.

Using information from potential clients that you have met, you may be able to write to those potential clients that were not willing to meet to offer a tailored service. An example of this would be as follows:

Dear Mr Jones,

Flexible working methods reduces laboratory capital investment.

From our previous telephone conversations, I recall that you will be shortly investing in a new laboratory complex as part of your R & D department upgrade. I thought therefore you would be interested to know how we were able to save 18% on the original budget for one of our client's laboratory projects.

By revisiting with our client, how they currently use their laboratory space, and introducing the sharing of key equipment, we were able to reduce the overall footprint of the new facility.

I believe the experience we have gained from this exercise would greatly benefit your business when reviewing your proposed laboratory facility. I attach a case study of our completed project and our client is happy for you to visit the facility in the near future if you wish.

I will contact you in the next few days to enquire if the idea of a visit, or explanatory meeting, might be of interest to you at this stage.

Yours sincerely

By focusing on how the actual service is to be delivered may require the process of delivery to be re-engineered. To keep ahead of the competition you will always need to be one step ahead. Perhaps there is an opportunity to bring in other professionals with unique services to the overall service package. This addition to the team may be the winning edge over the competition.

3.4   Features and benefits

All too often a professional service provider will tell the target client all about his firm and the services he has to offer. Hopefully, by identifying the client's needs, the professional can make this presentation more bespoke to the actual specific project. However, don't assume that the client can fully evaluate the benefits that may be derived from the various features you are putting forward about your services and firm.

When I put together documentation for a client to consider I always ask myself *so what?* after each statement. Here are a few examples to illustrate the point:

❑ **Feature:** We have been involved with over 20 office relocation projects in the past 5 years.

❑ **Benefit:** We understand, and therefore can address, current issues that arise, such as the need for layout flexibility, the latest thinking on incorporating the latest technology and how to obtain stakeholder involvement early in the process. This means we are able to propose office layouts that suit the current client's working methods and be flexible enough to adapt to future needs. Also by keeping the client's stakeholders involved, we get their co-operation early, thereby getting decisions made quicker. This all saves time and cost and results in a better-bespoke solution.

❑ **Feature:** We have, as a firm, won several awards for our designs, especially for their sustainable features.

❑ **Benefit:** This will enable the client to meet their current sustainability targets, improve their reputation and, above all, save on running costs and reduce their carbon foot print. This also has a wider benefit to the community.

In **Table 3.3**, there is a list of potential features that might be highlighted by the professional services provider and a list of potential benefits arising from those features.

3.5   Building trust

Selling professional services is more psychological and personal than other forms of selling. The client is exposed to more risk when buying a service as opposed to a product. With the absence of a product to

Table 3.3   Examples of features and benefits provided by a professional services firm

Examples of features and benefits	
Features	**Benefits**
We are a large firm	We are able to fully resource the project and take on board any additional work arising. This reduces the risk of delay and possible subsequent additional costs to the project
We are a small firm	Your project will be very important to us and we will have a dedicated director/partner running the assignment. This will make sure your needs will be accommodated as they arise
Our employees work within sector-specific teams	Whatever project you assign to us, we will have a team available that has the experience of the particular building type. This reduces the learning curve and gives the client the benefit of current good practice within the sector. This will also reduce costs, time and enhance the quality of the project to the appropriate standard
We are a national firm with offices throughout the United Kingdom	This will enable us to manage the project locally and with the benefit of local knowledge we will have experience of local contractors and suppliers. This will save money and be more efficient. Also we will be able to service your needs throughout the United Kingdom more economically. These savings will be passed on to the client
We have a value engineer within our firm	We are able to hold regular value engineering workshops at key stages within the pre-contract stage to check that the specifications are appropriate to the brief and use of the building. Invariably this results in cost savings which are passed on to the client
We employ an in-house visualiser	We are able to quickly create 3D visuals of the project and computer-generated walk through. This gives the client and his stakeholders a good representation of what is being proposed and the ability to make changes immediately rather than when on site, when modifications would be costly and could cause project delays

Continued

Table 3.3 *Continued*

Examples of features and benefits	
Features	**Benefits**
We are part of a benchmarking group	This gives us the opportunity to compare the various parts of the service with similar organisations throughout the United Kingdom. We are then able to address our weaker areas and build upon our strengths. This gives the client a better service and the comfort of knowing that we are always improving on how we operate and deliver our service. This makes us more efficient with savings being passed on to our clients

evaluate, the client is left to evaluate the 'seller'. In his evaluation of the seller, he must decide if he can trust him on the following:

❑ Are the seller's claims true (credible)?

❑ Will the seller deliver what he promises to deliver (reliable)?

❑ Will the seller be able to solve all the client's needs (capable)?

❑ Will the seller's organisation fit in with the clients' organisation and so make the project run smoothly (compatible)?

Building trust comes with time but being able to demonstrate you are capable, credible, compatible and reliable can help the process. Satisfy and communicate these attributes for the potential client and you stand a good chance of winning the assignment.

Capability

Most clients will question an organisation about their ability to undertake the assignment. The following are the elements to consider:

❑ **Track record.** There will be a need to show that you have undertaken several similar assignments. Project sheets, articles, client references and endorsements, visits to previous projects and presentations can demonstrate this.

❑ **Technical ability and insight to clients business and sector.** The client does not want to educate his consultants or contractors. He will need to be convinced that you are likely to know his key issues, know the current thinking in his sector and know his terminology. Also do you have the appropriate qualified people on your team?

❑ **Application.** It is one thing to have the track record and technical ability but you need to show that you have the expertise to adapt them to suit the targets specific needs and requirements. Perhaps also the leading edge thinking that can give added value to the project.

Credibility

Clients will be looking to see if there are any flaws in the organisations they are considering. They need to be reassured that they can believe all the claims that you will be making. So there is a need to do the following:

❑ Be reassuring and confident in delivering your message. You will come across as more believable if you believe in yourself and your ability. Do not give any reason for the client to doubt your ability. Rouse enthusiasm and confidence.

❑ Have facts and figures at hand or memorised to back your statements.

Reliability

Clients want to be assured that once they appoint the professional services firm, then they will meet their deadlines and deliver what they were going to deliver. They may have a resources problem and be over-stretched; the team that sold the firm might move on to the next sales pitch and hand over to the not-so-good team. Therefore the professional services provider will need to demonstrate that he is reliable by

❑ having third party endorsements to give the client comfort knowing that others have picked you before and you have rewarded them with doing a good job;

❑ having case studies showing that assignments were within budget, on time and to the desired quality.

Compatibility

You could be the best professional services firm in the market place but unless the client is convinced that he can work with you then you have problems in securing the assignment. The client wants to be sure that

❑ you really know and understand his business, his culture and his needs;

❑ you have a shared vision on how the project should be delivered;

❑ your team is able to work alongside the clients' team.

Rational and non-rational selection criteria

Of the above four trust-building factors, three are based on rational evaluation. These are capability, credibility and reliability. These are assessed by the examination of references, endorsements, accreditation, experience and qualifications of the team. Compatibility, however, is a non-rational evaluation. The potential client will develop a 'feel' or 'gut feeling' if he and his team will be able to work with the professional services firm.

Many of the professional services firms that will be selling to the potential client (your competitors) will probably all be featuring on the rational criteria. This is the easiest to demonstrate, especially if the firm has been religious in maintaining its records of its past work, developing case studies and obtaining client endorsements. What many of your competitors will not feature on, or not to a high level, are the non-rational components of trust – that is the compatibility angle.

It is this that makes clients more comfortable to work with people they have worked with before. It's a bit of 'better the devil you know.'

Therefore if you are able to shape your service and show to your potential client that you are compatible with his organisation you will have a good chance of scoring high on the evaluation front.

Most professional services firms will be describing to the client how it feels to work together in the future. The smarter professional services provider will try to give the potential client a sample of how this would be like. To do this you need to engage with the client while you are shaping the service delivery. Don't just disappear after the first meeting and then re-appear with a proposal (or worse sending it by post or e-mail). Give the potential client a sample of your thinking process and how you are building up a service delivery proposal. You can do this by going through an interaction process of the following:

❑ **Defining the needs.** Checking that your listening skills have captured all the points. Don't be afraid to ask clients for more detail or clarification.

❑ **Defining the problems.** Try to get the potential client to elaborate what is behind the needs. What is to be solved and who are the stakeholders who need to be satisfied?

❑ **Exploring the implications of the problem-solving.** Will there be other problems arising that need to be addressed?

Don't jump to solutions until you are sure (or reassured by the client) that you truly understand the problem. Why not have workshops between your team and the client's team? This not only helps to clarify the issues but also allows you to focus on the real dominant problem and also helps to build the relationship, rapport and trust.

3.6   Differentiating

Your competitors are probably approaching your target client on a regular basis. Therefore there is a need to stand out from the rest. You need to differentiate yourself and what you are offering from the competitors.

When you analyse your business, are you able to identify your unique selling point (USP)? Are you able to identify the following:

❑ What are the specific benefits from the service you are offering?

❑ Is your service unique in some way in that your competitors do not offer it in the same way?

❑ Is the service so special in its own right that it will cause the potential client to switch from existing suppliers to you?

Some people get confused about USP. USPs can be

❑ being the first;

❑ being a one off;

❑ market speciality;

❑ heritage (being in a specific market longest);

❑ being the latest in the field (to have designed the last Olympic stadium).

USP is not

❑ being the cheapest.

It is very difficult to differentiate on price. If your message is that you are cheaper than your competitors then you may undermine your aim to be different. Of course you may be 'cheaper' or shall we say of better value compared to competitors because

❑ you are multi-professional and so able to reduce overheads for the client's benefit;

❑ you are expert in the clients sector and so have no need to go through a learning phase;

❑ you have invested heavily in technology, which enables you to do tasks quicker and more efficiently, which gives the client savings.

The other key point is to find the attributes that clients like in their consultant, contractor or advisor. There may be surveys available or if not, instigate a survey (see Stage 6) and find out what your target clients value most. If you can identify these special attributes then package a key one, or several similar ones, and try to differentiate yourself.

Should you discover, for example, that your target might be very risk averse, package and present all the qualities such as the following that could make you the perfect choice:

❑ You have a specialist risk avoidance champion.

❑ You produce and regularly review a risk register on a wide variety of topics.

❑ You hold workshops with clients during the design and construction stages to make sure that the project is and stays within budget.

❑ You take stakeholder management seriously so that all the key people are on board with the project.

❑ You take seriously in-house quality assurance procedures.

The key in differentiation is not to be all things to all people. By all means change your pitch to suit your target client. Differentiate in a way that sets you apart from the competitors. The key is to make the client think you are a perfect fit to his culture approach and that you share the same values. You will need to find your winning edge (**see Fig 3.1**) to differentiate your offering.

During your research into the potential client you will discover what is important to him. Your initial meetings to discuss the client's

If the professional services provider does not offer added value then the client will select on price

If the professional services provider is able to provide benefits, then this gives him a possible winning edge over those who are offering a low price

Figure 3.1   Differentiate your service by offering added value through benefits.

needs will also have uncovered the reasons behind some of the needs. For example, your client contact wants to be able to check on the progress of the project at very regular intervals. You discover that this is because the person he reports to will ask for an updated report with no or little notice. Therefore, to tackle this you might be able to offer internet access for him to a special client extranet where he will, at any time, be able to log in and have a current status on all aspects of the project as updated by the team daily or weekly.

Therefore, knowing the reasons behind some of the client's needs will help to make the service offering more bespoke. Try to capture the real critical issues, the 'hot buttons'.

When delivering a message it becomes more powerful if it can be delivered within three concise statements. Politicians and professional speakers all know the power of three. Try to uncover and feature the 'hot buttons' for the target client. When you sell your offering, highlight the three key areas that go to the heart of addressing the client's needs. You are then well on the way to differentiating yourself and becoming more bespoke to the client needs. When in the briefing stage, why not ask your client to pick out the three most important issues. Talk around each issue and see how you can shape your service to address these issues.

If you then need to expand each of the three points, try to find three ways to describe the benefits to the client. The power of three is a forceful tool.

Differentiate through customer service

The ideal situation for the professional service firm is to acquire a level of competitive advantage that is sustainable over a length of time and has the client perceiving benefits over that time.

If your competitors can match any competitive advantage that you develop then your prime position is short lived. If you are able to differentiate yourselves and appear to give better benefits than your competitors, other than price, then you will be able to

❑ command a premium price for your service;

❑ achieve better client loyalty;

❑ sell more services.

So how can you improve your level of customer service so that it becomes a differentiating factor? Well, the following are some of the areas you could tackle:

❏ Change your culture so that it is client driven. The culture should be driven from the top down. This should not be just lip service to some nice-to-have goals.

❏ Invest in good people. Professional services firms 'sell' people, so your people need to be better than your competitors.

❏ Invest in training. Find out where there are weaknesses and train your people to excel.

❏ Have a positive rather than negative culture. A glass half-full approach rather than half empty.

❏ Appoint a client service champion.

❏ Try to integrate the client into your organisation and teams.

❏ Deliver your promises.

❏ Improve the quality and feel of all that the 'client touches'.

❏ Enhance all forms of communications.

❏ Strive to be the best.

❏ Empower your employees.

And above all,

❏ Keep on doing it and keep on improving what you are doing. Keep the client relationship fresh and do not let the relationship diminish.

Show you care

Research has shown that somewhere in the order of 68% of clients change their suppliers because they perceive that their suppliers become indifferent to their needs. It is therefore not good enough to give a good service; the client must perceive that they are getting a good service. The service needs to be better, faster than anticipated. Continuously caring for your client and making them aware that they are important to you will go a long way in retaining them.

3.7   Consider your strategy

At the conclusion of this stage, you hope to be able to communicate back to your target client that you are able to offer a bespoke service to suit his potential needs. It might be that the target client is still not in 'buying mode' and all that you are doing is putting yourself on his radar. You will need to find ways of keeping in touch and developing the relationship further and obtaining more intelligence about the opportunity identified.

Very often, you may be focused on a particular project, which continues to be delayed, and in the meantime another project for the same client comes out of the blue. If you haven't been keeping in touch you would not be 'in the right place at the right time.' The next stage explains how you keep yourself in the frame.

Understanding why clients might not want your services

There may be various reasons preventing you from building up rapport, or the potential client may seem negative about your firm and its services. Finding out what those reasons are can help break down the hurdles and help improve your chances to make winning proposals to future clients. The reasons why your potential client is not so forthcoming could be as follows:

❑ **Risk.** The client might feel vulnerable. He already has professional services suppliers and there is a potential risk in change. He might be uncomfortable about the value of the services being offered. He may be unable to establish if it is value for money and if there are sufficient benefits for the investment. Clients do not want to be seen to be making a mistake.

❑ **Suspicious.** Some clients may be suspicious if your claims are too good. Perhaps the way to overcome this is to have good references and by inviting your potential clients to visit your other projects and clients.

❑ **Don't want to make a decision.** If your offering is a major project or is very expensive (as perceived by the potential client), the client may defer making a decision. A way around this could be to break down your proposal into chunks so that the client can commit to a small investment, perhaps for feasibility, so he can sample if he is comfortable working with your firm.

Summary check list

By the end of Stage 3 you will have taken and reviewed your potential client's needs so that you can

❑ analyse the situation and decide if it is still worth pursuing the opportunity;

❑ investigate if you will be able to give enhanced customer value;

❑ identify a bespoke service with sufficient features and benefits to interest the potential client.

You will be able to review how you can build the relationship with the potential client and you will be able to communicate that you are

❑ capable,

❑ credible,

❑ reliable,

❑ compatible.

To make sure the potential client gives careful consideration to your offering, you will also have investigated what differentiates you from your competitors. This could be through

❑ providing a niche specialist service;

❑ the provision of unique benefits to the potential client;

❑ providing exceptional customer service.

Stage 4:
Communicating your availability and capability to the target clients

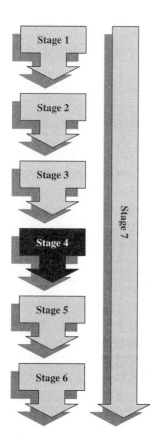

4.1 Communicating to target clients

4.2 Target clients in buying mode

4.3 Target clients not in buying mode

4.4 Raising your profile

4.5 Entertaining

4.6 Seminars

4.7 Writing articles

4.8 Public speaking

4.9 Memberships of organisations

4.10 Advertising

4.11 Exhibitions

Stage 4: Communicating your availability and capability to the target clients

4.1   Communicating to target clients

In this stage, we will look at communicating your availability to the following people:

❑ The target clients you have already met and have established their specific needs

❑ Target clients that you haven't met but are in communication with

In Stage 2 we looked at identifying the needs of your target clients and in Stage 3 we looked at making sure you were able to satisfy those needs. Now you must look at keeping in touch with your target clients so you are able to identify when the switch point occurs. That's when they come into the buying mode. Buying mode can also include their lead time to making a decision on how to proceed. A good salesperson will have kept in touch and be able to track any opportunities that may appear over the horizon.

From our earlier discussions with the target clients you should know

❑ the client's needs;

❑ the client's buying process;

❑ who the decision makers are and who the influencers are.

Your campaign will use this information to assist in securing new work.

4.2   Target clients in buying mode

Having established that a target client is in buying mode and having been able to find out their needs, you need, to have a campaign to move to the next stage.

Identify decision makers and influencers

During your initial meeting with the target client and perhaps through other conversations, say over the telephone, you have built up a client profile. Within this profile you have identified who the key decision makers and who the influencers are. You should know the following:

❑ Who actually will make the decision? This might not be the most senior person, who might sanction expenditure but rely on others to select the professional services provider.

❑ Who will the decision maker consult with before arriving at the final selection? Are there external advisors or internal people with a vested interest in the project?

❑ Are there any people who might not be affected by the decision but still are consulted or are able to influence the decision maker?

If possible, all these people need to be aware of your availability and suitability to the project. However, this can be a difficult area in that the persons you are dealing with may wish to give the impression that they are in charge and it is totally their decision as to who gets appointed. Therefore, they may not be willing to divulge who else is involved, and even if they do, they might not like the idea of you approaching other people. If this is the situation, you must get your target client contact totally on board and approach others within the target organisation, or external advisors, only with his blessing.

If this stage is mishandled, you run the risk of alienating your contact or losing the business totally.

In an ideal world, this situation could be overcome by doing sufficient research at the outset and getting to the decision maker. All too often, it's very easy to get to see someone lower down in the process and then it becomes very hard to bypass them to get to the right people. The other view is that often the decision makers will not be part of the selection process. Let's say, for example, a manufacturer needs a new warehouse. Our research might reveal that the following people are involved:

The managing director: Will actually sign off the project

The operations director: Will put the recommendation to the board

The operations manager: Will be doing all the work in identifying potential professional service suppliers, preparing a tender or brief and will be evaluating those pitching for the work

The finance director: Will be looking at the costs, cash flow and value for money

The procurement department: Will be looking at the contract of appointment and all the details

The warehouse manager: May have some influence in the selection process

---

Meeting the decision makers

Having built up a relationship with your target client contact, you are now able to meet with him and show him that you are able to assist him with his project. The meeting needs to demonstrate that you have listened to him in earlier conversations and meetings and give him the confidence that you can deliver what is required.

You now need to sell the idea to your contact that, to give him the best possible service, you still need to have more information from some of his colleagues. You need to sell him the idea that this will make him look good with his colleagues because

❑ the needs of his colleagues (other decision makers and influencers) will be taken into account and considered;

❏ he will be seen to be very thorough in his method of procuring the project;

❏ if all the necessary people are consulted through him, then there is less risk of the project being hijacked or killed off.

Once you have the client contact on board, you need to meet the key decision makers. Resist at this stage to present to them. At the moment you only have your client contact's view of the project. Some of the decision makers may have other views or hidden agendas. Try then to arrange a meeting with these decision makers. You will need to persuade each one that you are the natural choice for the project. Each person within the decision-making process will need to be courted differently. Obviously, there will be a consistent overall message, but within each discussion there will be emphasis on different aspects of the project and the service delivery.

When meeting the people within the decision process, don't rush to tell them about yourself and how you will deliver the project. Take some time to confirm their views and preferences. Don't assume from their job title that they may have specific views or interests. You will in time need to find some background on them, their past experience, relationships with other key people, views of the competition and any specific worries or concerns about the project.

Prior to the meeting you will have carried out some more research, probably through your original contact, in finding out the following:

❏ What is their role within the decision process?

❏ What role do they have within the business?

❏ Are there any known specific needs they have relating to the project?

Taking the warehouse example above, you may find you are explaining to the finance director how you have been able to control costs and bring the projects in within budget. The operations director may want to hear how you are able to help him become more efficient, or run his operations with less energy consumption or optimise his current layout to streamline his operations.

At each meeting it is important to develop relationships further. Also show that you are listening and taking note of their views and needs. Don't go and preach to them or present a scheme where they have had no input.

Meeting the influencers

These people might not have the final say but they may be able to lobby those who do. These people are usually the users of the facilities. So for a new hospital they would be the doctors and nurses, and in an office it would be the office workers. Don't overlook these people. It can't do you any harm to have them on your side, but it might make your job difficult to have them against you.

These people like to be kept informed. They want to know what the project will mean to them. How will it affect them? Further down the process you may need to present to them. A body of influencers may have one or two key people who tend to have more influence than the others. Find out from your client contact who these people are. Your success will be determined by your ability to win over as many people as possible. To do this, you need to reconfirm their needs and iron out, if you can, any conflicts.

When access is denied

Some clients will not permit you to speak to anyone in their organisation once the project has been identified and the selection process is underway. This is often the situation in public procurement projects. Ideally, in this situation, you have been tracking the project (in your pipeline) and been able to build some relationships prior to the project coming live.

Alternatively, you might be able to meet people involved in the selection process but they may be obliged to let all others bidding for the project know the questions you asked and the answers they gave. The trick here is to do the following:

❏ Be seen to be proactive and ask questions. But don't give good ideas away by asking questions which could alert your competitors.

❏ Try to meet the people rather than rely on questions and answers by e-mail. If you make a good impression during this initial stage, then you have an advantage over your competitors who will be seen not to be as keen as you. Clients like enthusiasm. Think about the key things you have bought and how it was sold to you. Enthusiasm is infectious. An indifferent salesperson can ruin the chances of securing the project.

❑ When meeting the client representatives on the pretext of a fact-finding mission, find the opportunity to 'sell' yourself and your firm. Do it in a subtle way.

❑ Always keep in contact, without being a pest.

4.3   Target clients not in buying mode

During this stage, you are also able to forge links with more than one person within the client organisation. You need to create the opportunity to be introduced. Having established the needs of the target client and their interests, there are many opportunities to further the relationship. You need to be careful not to irritate them; just keep in touch at such a level that the target client is comfortable with. I find it always a good idea to agree with the client on an approximate time in the future when to get back in touch, but never longer than a year. Perhaps when telephoning to enquire about the current situation, it is also useful to say to the target something like the following:

> *Thank you, Mr Smith, for giving me an update. It seems that nothing is going to happen with that project for at least 9 months. Would it be alright with you if I keep in contact and perhaps telephone again in about 6 months?*

You notice that this establishes the following:

❑ Agreement with the client on when to call again.

❑ Allows the client to vary the timing of the next call.

❑ Gets the client to agree that the next action is with you to follow up. Never, if possible, end the call with the action with the client. It will rarely happen.

❑ If you have established that something may happen in, say, 9 months, don't agree to call in 9 months as projects may be brought forward. All too often projects will be delayed, but you will find that as the project becomes more like a reality then your lapsed time before the next follow-up call will reduce, perhaps to 3 months and then to 1 month, etc.

Try to establish what will be the action that will cause the switch from the not buying to the buying mode. It could be one of the following:

❑ A start to a new financial year and the need to put budgets in place. Perhaps you could offer some free help to the client in establishing his budgets.

❑ A board decision. In which case, find out when the proposal will be submitted to the board and if there is a higher board that it needs to be referred to. Often in large international client organisations, a division may first give permission for a project to be put up before the group board. This is usual if large amounts of money are involved or if the project involves a strategic change to the business plan.

❑ Sometimes decisions are postponed until a new person has been appointed to a particular position, perhaps a new operations director.

❑ If the client is dependant on external funding, say, a government grant or lottery funds, then there will be a lead-in period for evaluation and agreement. Sometimes the project may be put on a reserved list for possible funding, should funds become available.

When in dialogue with the target client try to find out as much as possible. Many clients will not want to divulge much, especially if your relationship with them is new and untested.

With each conversation with the target client, try to find out a new fact about the project. It is always useful to get the client to identify the project, perhaps by an internal project name 'Project Jupiter', for example. This way when you call again in 6 months or so you can start the conversation with something like the following:

> *Mr Smith, when we last spoke some 6 months ago you mentioned to me that there was a potential project that may be coming up in a few months time. You were not able to tell me the details of the project but you referred to it as 'Project Jupiter'. Are you able to tell me how that is progressing?*

This kind of introduction has the following benefits:

❑ It reminds the target client that he did actually speak to you and this isn't a cold call.

❑ It allows the client to pick up the plot from where it was left with some kind of update.

❑ It tells the target client that you have a professional approach in that, you have followed up at the appropriate time, you have noted the potential project and you are developing the relationship in a non-invasive manner.

Should the client tell you that there was no movement on the project since the last call then again agree on the timing of the next call. Also don't forget to enquire if there are any other projects that may be coming along in the meantime.

During the intervening period you need to keep your name in front of the target client. Send him details of projects you have just completed which would be of interest to him. Be careful not to dilute your profile. For example, if you are chasing an industrial client who is looking to develop a large state-of-the-art distribution depot, then it may be unwise to send details of your latest social housing project. You need to build your profile to suit the client's needs. If you have established that the client wants someone who can deliver something innovative, then it is worth rewriting your project profile to emphasise the innovative features of the project.

During this stage don't expect to be given a job or even included on the tender list. There may be reasons as to why you have been excluded. Equally, be realistic that you may sometimes be included on the tender list, or given an enquiry, just to provide the client with a price check. On many occasions, clients will have their preferred suppliers and will sometimes come out to the market to check on their supplier's value for money. As long as you appreciate this and perhaps even in your preliminary fact-finding stage, identify this possibility, you can make the appropriate business decision if it is still worth pursuing the target. It can also be a situation of 'the devil you know' for the client.

Many years ago, a study revealed that a potential client would only consider a new supplier after, on average, seven calls. Therefore, if you have established that you will be calling your target client every 6 months then seven calls would be 3½; years. So it pays to raise your profile in the intervening period to improve your chances of being considered.

As mentioned previously, try to obtain new information at every opportunity, but don't irritate your contact, and if you notice resistance then back off. There is no harm in saying something like the following:

> Mr Smith, you mentioned to me that you cover the Midlands. We also work in the south east. Can you tell me who your opposite number is in that region? or
>
> Mr Smith, thank you for the update and I will call again in 6 months as agreed. In the meantime, are you aware of any other projects within your organisation, either in other divisions or other regions?

Once you have this information, act on it as soon as possible. Make that phone call and you could always introduce yourself by saying something along the following lines:

> *Mr Jones, I have just been speaking with your colleague Mr Smith in the Birmingham office about a particular project he may be progressing in the near future. During discussion, your name was mentioned and he suggested I give you a call. I was wondering if I could visit you to introduce myself, and how we may be of help to you.*

If Mr Jones hesitates, say something like 'It will only take 20 minutes of your time; I am sure you will find it beneficial.'

If Mr Jones refuses, then ask if it would be alright to send him some information in the post. If he says 'yes' and if you detect you can talk for a bit longer say something like

> *Mr Jones, is there anything at the moment which may be of particular interest to you? This will help me personalise the information I send you?*

You might be in luck, and a dialogue starts. If Mr Jones is not forthcoming then don't push your luck. Agree to send some information and then start the whole process of developing the relationship with him.

In time, you will be able to build up an organisation chart of the key people who may either give you work or influence the decision in your favour.

With all your contacts, you need to keep in touch with them in between the phone calls or meetings. That's why it is important to prioritise your efforts. Only, have a list of targets that you are able to service. Do not have a long list which you don't keep in contact with or is so daunting that you are put off from progressing.

Outlined below are some ways you can raise your profile and keep your business in the target client's mind.

4.4   Raising your profile

There is always a need to raise your profile. Even if you are currently overstretched and having to turn work away, there may be a time in the future when work starts to dry up. In some ways, the best time to start raising your profile is when you are busy. In busy times you will probably have some funds available to try out different methods of raising your profile.

In Stage 1, we examined what kind of client you wanted to work for and within which sectors. Bearing this in mind will steer you into what activity you might perceive being useful to raise your profile to attract the kind of clients you want.

There are many ways you can raise your profile and they include networking, entertaining, writing articles, public speaking, memberships of organisations, advertising and taking part in exhibitions. Many of these activities could be classified as marketing, but they are an integral part of selling your services.

Networking

If you want to be prosperous for a year, grow grain.

If you want to be prosperous for 10 years, grow trees.

If you want to be prosperous for a lifetime, grow people.

Networking is all about mixing with people, creating relationships and helping them achieve their goals, and in turn they will help achieve yours. The people you meet will have many contacts of their own, some of which may be useful to you. Networking is about tapping into these relationships.

Where to network

There will be many opportunities for the professional businessman to attend functions, some of which may be specific networking functions. A quick search on the internet will locate these specific networking clubs and events. When picking the functions, consider the following:

❑ The catchment area for the function. Is it local, regional or national?

❑ Who is likely to attend? The website might tell you the membership or likely guest list. If not, speak to the organiser.

❑ The number of attendees.

❑ The frequency of meetings.

❑ Who is sponsoring the event? This might give you a clue in respect of who might attend.

❑ Cost (money and your time).

If, after these initial considerations, the event looks as if it could expose you to other like-minded people, then it's worth sampling an event. If there is a membership joining fee, then ask if you can first attend as a guest so you can evaluate the event.

Of course, there may be networking opportunities at almost all types of gatherings, so always be prepared.

Prepare for the event

It is worth preparing for the event, and some of the things you could do before attending are as follows:

❑ Make sure you stock up with business cards.

❑ Take your own delegate badge rather than use the ones that are handed out. This will make you stand out and convey the information you want to convey rather than the information the organisers want to put on the badge.

❑ Ask the sponsor for a delegate list. Read through and mark up those attendees you would like to meet. Look them up on the internet, do some research.

Working the room

Networking is hard work. Working the room for maximum benefit can be very rewarding. It is so easy to attend a function and get stuck with someone and miss out on a whole load of potential contacts. To get the most out of the room you should do the following:

❑ Arrive early. It is much easier to begin to network when there are a few people there rather than a full room.

❑ If you want to join a group of people who are already in discussion then just walk up to one side and listen to the conversation. Try to make eye contact with one of the people; they might then turn to you and introduce you to the group. If they do not then wait for a pause and hold out your hand and introduce yourself. If you have been listening to the conversation, you might be able to pass comment on the topic and that way rapidly join in the group.

❑ Quickly evaluate if the people in the group are going to be useful contacts. If not, then move on by saying something like 'Excuse me, I must mingle'.

Introductions

First impressions are very important. Although you will have a chance on building on that first impression, it is important to make it a strong and positive one. You can do this by following these tips:

❑ Wear the appropriate clothes, look the part of the professional person you are.

❑ Introduce yourself in a clear interesting manner.

❑ When you are asked what you do, have an interesting answer prepared which gives an idea of what you actually do rather than just your job title.

❑ When attending an event, have a few ideas for 'small talk', listen to the news and pick out appropriate items which might come up in discussion, or interesting stories you could introduce to the group. Perhaps have a few set lines which you could use as conversation generators. Listen to how others do this and see which techniques would suit your style.

❑ In turn, you must listen to others and appear interested. Take the opportunity to learn about them and develop rapport.

Whatever you do, don't do the following:

❏ Hand out your business cards to everybody in the hope that you might strike lucky. Wait for a rapport to build up and then hand over a card if it's appropriate.

❏ Collect a meaningless collection of business cards without building some sort of rapport with the people.

❏ Try to make a 'sale' at the first encounter.

❏ Be too pushy. Gauge the situation and see where it goes.

The follow-up to a networking event

There is little point in attending networking functions and not following up. You need to prioritise and take appropriate action. To prepare for this you should do the following:

❏ Have a system of collecting business cards at the event. Have one pocket for your cards. These should be kept in a card holder so they are fresh and not dog-eared when handed over. Have another pocket to put the cards you collect. Use a third pocket to put those cards where there is a need to have a quick follow-up.

❏ When you collect the cards, you can make notes on them while talking to the contact. It shows you are taking an interest and intend to follow up. Soon after you leave, put a note on each card with the date and place you met the person and maybe a note of interest or action.

❏ Load up the information onto a database with some action or note of the conversation.

❏ If you have been given a card by someone you know will not be of interest, then discard the card rather than spending time loading up onto your database.

❏ Follow up with those contacts you developed a rapport with. Send them an e-mail saying it was good to meet them and suggest ways for how your new relationship could develop. Perhaps, give them some new contacts or arrange to introduce them to someone else.

The more you do for others, the more chance of someone doing something for you.

Make yourself a useful contact

The key to good networking is that you help others as well as ask them to help you. The more you help them, the more useful they see you are and hopefully will reciprocate. Therefore try to

❑ be an interface for the collection and distribution of information;

❑ share leads and opportunities with others (not with rivals);

❑ connect people with others;

❑ maintain regular contact;

❑ find opportunities and reasons to meet again to develop the relationship.

Creating networking opportunities

Once you have developed your own network you might find it useful to hold a small breakfast or lunch meeting which could

❑ be themed;

❑ have attendees who could be useful to each other;

❑ be just your organisation meeting another organisation.

Before you know it, you will be at the hub of a network of people.

Develop your network

Your network needs to be made a strong one, so take time out of your busy schedule to

❑ review your network every week and telephone one or two people who you have not talked to for sometime (no longer than 3 months);

❑ send out notes, newsletters etc. to keep your name on their radar;

❑ send clippings and articles to people if it relates to them or their business sector and include a short note;

❑ take every opportunity to develop the relationship.

4.5 Entertaining

A lot of money can be wasted on entertaining if it is not approached with a focused mind. Consider the various levels of entertaining and the merits of each.

❑ **Informal get together after a business meeting:** This is ideal to get to know the client better. It gives you both some social interaction away from the project. This could be a drink or light lunch after the meeting.

❑ **Ad hoc lunch:** This is pre-organised and may well have a specific purpose to find out some information or just to keep the regular contact going. Ask the client to select the venue so they are comfortable with the level of entertainment and time taken up during the middle part of the day. A good option can be to go for a buffet type of meal where you can pace the time taken and not have to wait a long time to be served.

❑ **Themed lunch:** Invite several clients with a common interest to discuss key issues with their peers. Perhaps invite a key individual who can start the lunch with a short presentation on a current topic. The clients will then see you as a facilitator of best practice.

❑ **Invitation to formal dinner:** Often 'black-ties' function. Make sure it's an interesting function or is of particular relevance to the client.

❑ **Invitation to a sporting event (to participate):** A common choice is golf. How about trying something unusual like fly fishing with tuition or air balloon ride?

❑ **Invitation to a sporting event (to watch):** Common ones are rugby and football. Again, try something unusual and appealing.

Often the most important clients are the busiest and are usually inundated with invitations. What does go down well is the opportunity to involve the clients' spouses or, even better, their children. Hire a small cinema and arrange a private screening of a new film just released; invite children to football or rugby matches where they get to meet the players and tour the trophy room. Go to the theatre and have a backstage tour and meet the stars. Children really like these kind of events. If you succeed in getting your clients to these events with their children, they are usually very grateful.

4.6   Seminars

Seminars are a good vehicle for networking and for raising your profile. There will always be opportunities to

❑ attend seminars at conferences;

❑ deliver a seminar at a conference;

❑ attend other peoples' seminars;

❑ run your own seminar.

Attend seminars at conferences

Focus on the conferences that are targeting the sector you wish to expand into. You should look at the marketing literature for the conference to see who the likely attendees will be. Preferably you want a high proportion of level 2 people. Alternatively, if it is an annual event, contact the organisers for a breakdown of attendees from previous conferences.

If you are a delegate then you will normally be given a delegate list in your joining pack. This could be very useful in identifying people you want to speak to during the conference. This has the added benefit of possibly targeting people who work for a company who have a no-names policy when contacting them.

Use the information from the conference seminars/workshops to add to your growing knowledge of the sector. You may also pick up on the current themes and issues within the sector. Use this information, without breaching copyright, to update your website, literature and pitches to show you are at the cutting edge.

After the conference, try to meet up with some of the conference speakers (if they are not competitors) and see if there are opportunities to work together.

Deliver a seminar at a conference

This is a great opportunity to promote yourself and your organisation. The conference organisers will do all the marketing and after the conference you could pursue key people from the delegate list.

As in all seminars the delegates want to learn, and they don't want a sales pitch. So resist the obvious self-promotion, just badge each PowerPoint slide with your logo and make sure all your contact details are in the delegate joining pack.

Attending other people's seminars

Be very selective on what seminars you pick. Try to establish the following points:

❑ Attendance levels.

❑ Seniority of people who will be attending (will they be level 2?).

❑ Will the seminar attract likely potential clients?

You will soon establish which organisations hold the best seminars from the point of view of networking opportunities. Don't forget to always get delegate lists. It's often the delegates who don't turn up that you need to follow up on.

Running your own seminar

There is a lot of preparatory work to do before running your own seminar and it can be expensive, especially if you need to hire rooms and provide refreshments. However, if you are able to get the right people attending, and follow up, then it is a good way of developing relationships with potential clients. Don't forget to also invite existing and past clients. They will be able to endorse your services and it helps keeping in touch with them as part of your ongoing strategy.

Listed below are a few points to consider if you chose to run your own seminar:

❑ Give as much advance notice of the seminar to your potential attendees so that they can put it in their diaries.

❑ Send out the invitations about 8 weeks before the event and chase up 4 weeks before by telephone if you haven't received a reply.

❑ When picking a date for the seminar stay clear of school holidays. Also avoid December as you will be competing with many social events. The best month is November as few people take holidays in that month. The best day of the week is a Thursday. Avoid Mondays and Fridays.

On the day, make sure you have sufficient numbers of your team available to mingle with attendees. At the end of the seminar, your team should meet up for a debrief and decide on a strategy for following up on any good opportunities that may have been uncovered. All the leads should be entered into your leads pipeline and then regularly monitored and the status updated.

Consider also those people who did not attend. For those who said they would attend and didn't, send them some seminar notes. Also contact those who declined the invitation and offer to send them some notes or offer to meet them to discuss the seminar topic.

For some clients, and potential clients, it might be a good idea to offer to do the seminar (or a shortened version) at their offices for the benefit of their team.

4.7 Writing articles

This is a very useful way of getting yourself and your business known. It is obviously important to write articles that will appear in publications that your target clients will be reading. Also get the publishers to reprint the article so that you can send copies to the appropriate clients and targets you are pursuing.

To make the most of writing articles you should do the following:

❑ List all the publications that your clients might read.

❑ List all the publications which might be interested in what you do.

❑ Find out who are the commissioning editors or features editors.

❑ Find out the type of articles the publication produces. Is it illustrated? Is it learned?

Once you have done this, have a brainstorming session and come up with some ideas that might be of interest. To stand a good chance it must be

❑ applicable to the publication;

❑ of interest to the readership;

❑ topical.

Look at the publication website and see if there is a section on writing for the publication. Some publications will say that they have retained writers and will not be looking for others. Some publications will welcome contributions and will even give advice on how to present your ideas.

You can approach getting articles into the press by providing the journalist with either a complete polished article or the appropriate information for the journalist to write his own copy.

Before you commit too much time on the project, give the appropriate person a call within the publication and sound them out on your idea. If there is a glimmer of interest offer to send them a proposal for their consideration.

Some publications might pay for an article, usually when commissioned. Speculative articles are not usually paid as it is seen as a marketing activity to raise your profile through their publication. You will also stand more chance if you are offering the article free. Irrespective of whether the article is free or paid for, it will need to be professionally written.

It is always best to send the commissioning editors a synopsis of the article covering the following:

- ❑ Main topic
- ❑ Main elements to be covered
- ❑ Approximate number of words
- ❑ If it will be illustrated
- ❑ How long it will be to deliver the article from being given the go ahead

When preparing a proposal, bear in mind what the usual length of the articles is. There is no point offering a five-page article if the normal length is one or two pages.

Also look, if possible, at back issues or archive material to see if your proposed topic has been recently covered. There will be little interest if the same topic was featured recently.

When you have done all this it would be a good idea to contact the commissioning editor and tell him you will be e-mailing the synopsis and that it was a follow-up from a previous conversation. Some publications are inundated with unsolicited material and your article might go unnoticed for a long time because of other pressures on the editor.

If you are not able to write articles, get to know the news or features editors on the key publications your target clients will be reading. If you become an expert, the editor will be in contact for a quote to be placed in a piece he is putting together. The more available you are to provide him with a copy, the more he will come back to you in the future.

If you obtain a list of forthcoming features, then you can contact the editor to say you might have an interesting point to make or have a good example of a project that could be used as a case study to illustrate the point being made within the article. If you persevere and build up a rapport, then you could become a regular contributor.

Perhaps write or e-mail the features editor along the following lines:

Dear Mr Harris,

I am aware that you will be dedicating your April issue to the pharmaceutical sector. Therefore I would like to bring to your attention the recent project undertaken for Pharmiagric Industries by Wren & Barry Architects.

Your readership will find our recent laboratory project of particular interest because

❑ *it incorporates the latest UK and USA thinking on flexible laboratory design for the pharmaceutical industry;*

❑ *sustainable features have been used, which has produced dramatic energy savings and reduced the clients carbon footprint;*

❑ *the project was designed incorporating some innovative design features.*

We would be pleased to provide you with a case study highlighting all the relevant facts and figures. Alternatively, our in-house experts are available to provide quotes or comment on various aspects of laboratory design within the pharmaceutical industry.

I will give you a call in a few days' time to see if this may be of interest to you.

Yours sincerely

From this type of letter you have been able to do the following:

❑ Offer a case study for inclusion within the forthcoming feature.

❑ Offer facts and figures which could be incorporated within the journalist's own words.

❑ Offer some expert to be available to provide quotes and comments. Journalists like to have a directory of experts they can call upon at short notice to contribute to their copy.

❑ Leave the action with yourself. Journalists are very busy people often working to short deadlines. By telephoning as a follow-up, you are able to make sure that the journalist has seen your letter or e-mail and to obtain some feedback.

4.8   Public speaking

Obviously, public speaking would be applicable to those who are able to deliver a good speech and have something of interest to say. But this needs to be directed to the right audience. Speaking to your peer group will not get you much work. Talking at seminars is a useful way to start on the speaking circuit.

Try to approach event organisers within your target sectors and make yourself known, and also your availability to speak.

4.9   Memberships of organisations

You need to be selective when joining organisations for the purpose of winning work. You need to satisfy yourself that

❑ the membership contains potential clients or contacts that can introduce you to potential clients;

❑ the organisation actually holds functions at regular intervals and at the right time and location to allow you to network.

If the organisation is useful to you for networking and raising your profile, then try to become involved within it. Perhaps try getting elected onto the committee. That could open doors for you and will also look good on your CV, which is also incorporated within your proposals and tenders.

4.10   Advertising

Many marketing managers have an advertising budget but not all of them use it wisely. Advertising, if used well, can be very powerful.

Advertising on an ad hoc basis is a drain on resources and its benefits are hard to evaluate.

Therefore if you are going down the advertising route, it should be part of an overall strategy, and you should have a plan. Consider the following questions:

❏ What is the message I want to convey?

❏ To whom do I want to convey that message?

❏ Where is the best place to advertise to convey that message so that my target audience can see the message?

If you are targeting a particular sector then research what specialist publications exist that your target audience will be reading. Have a short list and contact the advertising editor at the publication and ask some questions which may include the following:

❏ What are the readership figures? They should have a breakdown of numbers which has been audited and verified. These figures include numbers who subscribe to the publication or receive it free. Also there will be an additional readership if the publication is passed around the workplace.

❏ Find out the frequency of publication.

❏ Ask for the rate card and ask for discounts.

❏ Enquire about any features that may be coming up, which would be appropriate to your advert.

❏ Consider advertorial (paid for article about you or your work).

❏ Consider sponsoring a feature or publication.

❏ Would a smaller advert over several issues be more beneficial than one big hit?

If you are going to invest in advertising, then do get professional assistance if you don't have the skills internally. You will need someone to help convey the message, possibly a copy writer. You want someone who can do the graphics to give impact to the message and, if possible, you need to test your ideas out on someone from the target audience.

Images need to be brilliant and if you are featuring your projects then have them professionally photographed; you can share the cost with your client and other members of the professional team who will also want good images. If you are going to use library shots then don't forget the cost of copyright, which can be expensive.

When considering advertising don't get carried away. Sit back and ask yourself if this is money well spent or can you spend it better doing something else?

You might be able to come up with a deal with the publication that if you take a significant advert then they would do an editorial piece about you; in which case, try to have the advert in a different part of the publication or in another issue. That way the readers might not link the advert with the editorial and the editorial might appear unsolicited.

Sometimes you can use the publications to your advantage without paying for an advert. You could do the following:

❑ Submit news about your projects – project won, completed, etc. If you can, submit images or visuals (see Stage 7 regarding press releases).

❑ Submit news on people joining your company.

❑ Even job adverts will tell readers you are busy and successful.

If your budget can provide for some public relations advice then obtain some. Ask colleagues whom they use. Don't rush to pick someone. Do some research and perhaps ask a few to pitch for your work.

It is worth mentioning here something I call 'blackmail marketing'. This is where a client of yours has given your name as one of his suppliers to a publication. The publication has offered to do an editorial on your client that will be paid for by supporting advertising by the supply chain. I have always tried to resist these. These publications are seen as being full of advertorial and usually read by your competitors rather than future potential clients. Look at past issues and evaluate the benefit. Sometimes if you are a major supplier, then it is hard to say 'no' and risk offending your client.

4.11   Exhibitions

Exhibitions can be extremely costly not just in the cost of the exhibition stand but also the manning, travel and subsistence costs. Virtually every sector within the building industry will have exhibitions dedicated to it.

If exhibitions feature in your marketing strategy then visit exhibitions in one year with a view to exhibiting the following year. During your visit make a note of the following:

❑ Which are the 'hot spots' and where are the dead areas at the venue?

❑ Is it an exhibition linked to a conference and will delegates only tour the exhibitions during refreshment breaks? If this is the case, perhaps see if locations near refreshment areas do better than other areas.

❑ If the exhibition is linked to a conference why not spend the effort in taking part in one of the talks or seminars. If you speak to the conference organisers (usually an association and separate from the exhibition organisers) well in advance, offer them some topical subjects where you could provide speakers. Alternatively, you could sponsor a talk or seminar and be associated to the topic.

❑ If the exhibition is good, then it will be hard to get the best spots. These are usually booked a year, if not more, ahead. Why not ask the exhibition organisers to let you know if any of the prime locations do not get snapped up?

You can make exhibitions work for you by arranging clients to meet you there; mailshot potential clients to tell them you will be there and they will be welcome if they visited you. You could make use of the venue, especially if outside your normal territory to get some client visits in or invite potential clients to a themed lunch, at or near the venue.

Some points to consider should you decide to exhibit are as follows:

❑ Make the stand look professional. Use professional stand designers and use good images and few words.

❑ Remember, space below table height is useless for words. Place your key message at eye level.

❑ Incorporate something to entice the visitors onto the stand – a game or challenge, moving images on a large plasma screen or interesting exhibits.

❑ Consider providing refreshments for visitors who stay to chat.

❑ Record visitors who expressed some interest so you can follow up.

❑ Many exhibitors offer entry into a prize draw if the visitors hand over a business card. This can be useful to add to your database and for following up.

❑ If you are manning the stand with a colleague then don't talk to each other and ignore visitors.

❑ Some exhibition organisers have a system where exhibitors can scan barcodes from delegate badges.

At the end of the exhibition take some time to review the success, or otherwise, of the event. Did you get useful contacts? Did you follow up? Was there any possible work coming out from the exhibition? Would you attend next year?

Alternatively, would it have been as useful just to attend the exhibition and network during the seminar and workshop breaks?

Summary checklist

By the end of Stage 4 you will have embarked on a strategy of communicating your availability and capability to the target potential client. You will have

❑ identified when the client will be likely to appoint a professional service provider;

❑ be aware when the potential client switches to buying mode;

❑ have identified and met the decision makers and key influencers.

For those potential clients not yet in buying mode, but have future plans which are of interest to you, you will

❑ keep in regular contact to track the progress of the opportunity;

❑ send the potential target client updates, newsletter and e-mails to keep your name in his mind.

During this stage you will raise your profile within the community by

❑ networking;

❑ entertaining;

❑ possibly writing articles;

❑ public speaking;

❑ joining appropriate organisations.

You may also wish to consider the following to further raise your profile:

❑ Advertising

❑ Attending exhibitions

Stage 5: Proposals, tenders and pitching

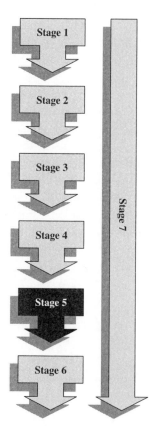

5.1 Proposals

5.2 Selling yourself and your proposal

5.3 Pitching for work

5.4 The selection process – direct with the client

5.5 Selection process – through and with a contractor

5.6 The trend for competitive proposals

5.7 Expressions of interest

5.8 Pre-qualifications

5.9 Tendering

5.10 Using CVs

5.11 Monitoring progress of the tender
or proposal

5.12 Post-tender interview

5.13 Negotiation

Stage 5: Proposals, tenders and pitching

5.1   Proposals

Preparing a proposal for a client is an important step in one of the last stages of the selling process. The need for a proposal might have been at the request of the client or at the suggestion of the professional services provider.

If you have been talking to the client for some time, you may have been able to identify a need that the client hadn't properly addressed. After establishing the client's requirements, you will need to create an opportunity to present your solution to the client. You do this before he decides to go to the market place with a formal request for solutions from various professional services providers.

Therefore, to keep up the dialogue with the client, you will offer to present him with a proposal that will focus on his problems and advise on how you are able to solve these issues. This gives you an opportunity to create an edge for yourself by presenting new ideas and solutions before your competitors are aware of the opportunity.

Create opportunities to build relationships during the proposal stage

Before rushing into preparing a proposal, you may wish to consider the following:

❑ Organising a visit for your potential client to a project of yours. This will allow your past client to endorse your services and also use the project to help clarify the potential client's needs.

❑ Organising a visit to your offices so that you can show off the uniqueness of your firm and your team. Take this opportunity to introduce other members of your firm and build up the relationship ties.

❑ Holding a workshop with the client's team, which will tease out the various stakeholders' needs and help develop even further inter-team relationships. Try to introduce lots of ideas at this workshop because it's important to be able to dismiss lines of thought as well as home in on the more popular and appropriate approaches. Use the session to show you are able to think outside the box and explore various solutions to the assignment.

By undertaking the above, you are able to continue preparing a proposal, which you are confident will be well received because it has been researched and tested with the client's team.

Keep the prospective client involved

To increase your chances of securing the assignment and hitting all the right buttons you need to keep the potential client involved during the proposal preparation stage. You should try to do the following:

❑ Obtain the potential client's approval to contact him during the process to clarify points.

❑ Visit key people within the client's organisation to go through various elements of the proposal (at a draft stage) so that they have signed up to your ultimate proposal. Try to make them think that they had come up with some of the thinking and ideas.

❑ Issue a draft proposal and arrange to meet the client for a discussion. During the whole preparation stage, encourage comments.

The proposal content

Each professional firm will have developed their own style for writing and presenting a proposal. The following should be the key elements:

❑ **An executive summary.** Condense the proposal into a series of clear, unambiguous paragraphs, which allow senior management to get a feel of the key issues of the proposal and summarise the key areas of the proposal.

❏ **Introduction.** This sets out the brief history leading up to the proposal, the outline of the assignment and how it was developed.

❏ **Putting the assignment in context.** This section will address the issues that the potential client is facing. In this section you need to demonstrate your knowledge of the client's business and markets, and his current business needs.

❏ **The solution to the client's needs and your ability to provide the service to resolve those needs.** You will need to outline any options you have considered. You may discard some options (with reasons) and describe any remaining options and provide recommendations. Any recommendations will have been tested during the preparation of the proposal with the potential client team. This section will need to demonstrate value for money, benefits, your strengths and uniqueness to provide the solution. Also, the solution will include programmes and costings where appropriate.

❏ **The delivery team.** This should be a bespoke team with CVs. Do not put in generic company structures here. Show how the delivery team will be part of the client team and lines of communication, and what its responsibility will be.

❏ **Appendix.** Use appendices to include the agreed brief, research, outcomes from workshops and visits, background information of your firm with appropriate examples of similar client endorsements, etc.

Depending on the size of the assignment, your likely chances of success and current commitments will determine the amount of effort you are going to put into a proposal. If you have arrived at this stage by going through the other earlier stages then you should be totally committed to producing the best proposal you can, given the time and resource constraints. If, however, this opportunity has bypassed some of the earlier stages and you are not sure about the client, then you might be hesitant in how much effort to put in. If the proposal means that you are, in fact, developing a brief and embarking on feasibility work you might be able to charge a fee. This is obviously up to your view of effort versus potential reward.

Approach to fees

You will obviously need to address the whole issue of fees for implementing what is outlined in the proposal. You will need to consider how to address this issue:

❑ Provide an indicative fee or firm commitment. This could be linked to how firm the proposal is and how many options there are.

❑ Include the fees in the proposal or leave it out. It is usually better to leave the fees out of the proposal document and only discuss these with the appropriate person in the client team. You don't want your fees to be broadcast throughout the clients' organisation, as you may not have control on the internal circulation of the proposal document.

❑ It is also best to leave the discussion on fees until there is a firm definition of the scope of works and the services being provided. You could see the proposal document being stage one, with a stage two being a submission on fees.

❑ If the assignment is straightforward and once you have feedback on the initial proposal, then expect to discuss fees.

5.2   Selling yourself and your proposal

Selling professional services is different from selling products. We do not have an object to describe or sell. The process is therefore more personal. Your potential clients have to evaluate you. They are not only looking at your experience but they will be making a judgement on their ability to work with you.

Quite often in selling professional services the seller is also the deliverer. Many clients want to meet the team that will deliver the project rather than the sales team. Up to this stage you might have been able to have a 'front' man, a charismatic person who can be convincing, sincere and be seen as suitable to the client's needs, culture and approach. All these points will become part of the professional's toolkit.

Now the time has come to be project specific, to tell the client how you are better than your competitors, on the basis of his criteria you found out during the preliminary stages.

Most sales people are extroverts and are happy to sell their firm's services. These people tend to focus on the overall big picture, but are not happy looking at the detail. The delivery team will, in contrast, be happy with the detail. The delivery team will also find it difficult to sell themselves, often being modest and not liking to boast about their talents.

Given the lack of a physical product, the potential client has no option but to evaluate the personal characteristics of the team as well as their expertise and past achievements.

In some ways, the professional client knows that there are risks involved. He knows that the professionals probably know more than him (unless the client has hired in an expert). The client is afraid of failure and often there is a lot of money involved. Of course, the client will be looking at the cost of the delivery of the service and looking at the team's credentials, and he may take up references. A big factor in the evaluation process is still about people.

To increase the chances of success, the selling professional needs to embrace the fact that people choose people. Therefore during the final stage of securing the project, the successful service firm will need to build trust. Hopefully, during the previous stages the relationship has been building up and the credibility, compatibility and capability attributes have been established. Now all the preparation has to be focused on a particular project with a bespoke team.

The client will realise that at the outset there will be information lacking in his brief, there might be insufficient money to deliver the ideal project and all too often not enough time to properly develop all the options and arrive at the perfect project. Therefore the client will need to trust the delivery team to take him through the process with the minimum exposure to risk.

5.3   Pitching for work

Purpose of the pitch

The purpose of the pitch will depend on the following:

❑ The current relationship. Is it an introduction or an elaboration, old or new contact?

❑ Are you pitching for a specific project or to get on to some short list or framework agreement?

❑ Is it the client, their advisors or other consultants you are hoping to work with?

The fundamental point about a pitch is to achieve your objective. This could be to win work or to build a relationship with others, which can then help you win work. There is no point in doing a pitch without a clear objective and outcome.

The pitch may occur at any stage of the process or there may be several as the selection process proceeds.

Plan of action

Having established the purpose of the pitch, there needs to be a plan of action. A person must be put in charge of the pitch. This person could be the key account person, the in-house business development manager or someone else drafted in. The person would have to have a degree of authority or have the backing of someone who can make things happen so that the process goes to programme.

A plan of action needs to be drawn up to make sure nothing is left out and the pitch team has the best chance of winning the work. The initial list for consideration would include the following:

❑ The brief

❑ The timetable leading up to the pitch

❑ The participants

❑ The audience

❑ Content of the pitch

❑ Format of delivery, use of technology

❑ Questions you don't want to be asked

❑ The venue

❑ Rehearsals

The brief

The foundation to success is a good brief. The client account manager should have built up a client profile, his needs, wish list, pet hates, etc. There may well be a formal invitation to the pitch and this invitation might be very prescriptive on content, time allowed, who must be present and the format to be used. As a minimum, these requirements must be fulfilled. All too often I see presentations where the people pitching just deliver what they think the client should know, rather than what the client wants to know.

Good preparation and intelligence gathering leading up to the pitch should have uncovered additional information, which would be advantageous. It is important to know the reasons behind the requirements. Was there a bad previous experience? Is there a hidden agenda?

So, building up the briefing document is critical. Also, background information about the client organisation is useful especially if you are able to weave in some client facts to illustrate a point in your pitch. This shows the client you have taken some time to understand his needs.

Timetable leading up to the pitch

If you have followed the stages in this book you will already

❑ have been aware that the opportunity to pitch might occur;

❑ have prompted the pitch, or even requested the opportunity to pitch;

❑ have some background information on the client and may have already met and taken a brief.

The pitch manager must draw up a timetable from the outset. The timetable must include all the milestone events and the development of the deliverables. An early meeting should be called for with all the key people involved (see **Table 5.1** for events leading up to the pitch).

The participants

Who attends and presents to the client is an important factor. They need

❏ to be there for a purpose;

❏ to deliver part of the presentation;

❏ if possible, to be charismatic, enthusiastic, sincere and knowledge-able.

This last point is a difficult one. Clients don't want to see the 'salesman' who is good with the presentation but not part of the actual delivery. Often it is best to include the person who will deliver the project. The client wants to know whom he will be dealing with. He needs to build some rapport, be able to trust them.

You also do not want too many people; you will be guided by the available time and information to be covered. Equally you don't want three or four people attending and just one person doing the presentation and fielding all the questions. The client wants to know he is getting a team, not just one person, who in fact might not be involved in the actual delivery of the project.

There needs to be a chemistry among the presenting team members, and if they have worked together before on a similar project, then all the better.

It is worth mentioning here the need to have the delivery team presenting to the client. Some professional services firms, especially the larger ones, may be tempted to have their best people (their 'A' team) present and then, when the job is secured, hand over to another team for delivery. This can upset the client relationship.

Let's look at another approach using three potential teams:

❏ **The 'A' team.** This is the best team you can put together for the pitch. They have the relevant experience and can easily deliver the project.

Table 5.1 Key events leading up to the pitch delivery

Key events leading up to the pitch delivery		
Pitch stage	**Pitch manager action**	**Delivery team action**
Stage One	❑ Read documentation ❑ Decide on delivery team ❑ Decide on any support team ❑ Initial brainstorm with team ❑ Decide on key issues and needs ❑ Research background on client ❑ List and allocate tasks ❑ Prepare internal timetable ❑ Start dialogue with client (find out competition, etc.) ❑ Prepare competitor analysis	❑ Read documentation ❑ Feed questions to pitch manager ❑ Attend initial brainstorm ❑ Take on board tasks ❑ Research client
Stage Two	❑ Continue dialogue with client ❑ Pass any feedback from client to delivery team ❑ Confirm venue and perhaps visit, check equipment available ❑ Confirm pitch structure with delivery team	❑ Any questions for the client, pass onto pitch manager ❑ Start to develop individual presentations
Stage Three	❑ Continue dialogue with client, test out any ideas ❑ Manage first rehearsal ❑ Invite attendees to rehearsal (external to delivery team) ❑ Support team also attend ❑ Take on board feedback ❑ First draft of support documentation ❑ Double check whether first draft addresses all the issues raised by client and incorporates added value	❑ First rehearsal run through ❑ Contribute first draft to support documentation ❑ Identify features and benefits to include within proposals ❑ Read first draft and pass comments back to pitch manager

Table 5.1 *Continued*

	Key events leading up to the pitch delivery	
Pitch stage	Pitch manager action	Delivery team action
Stage Four	❏ Manage second rehearsal ❏ Take note of feedback from attendees ❏ Start finalising support documentation ❏ Finalise presentation	❏ Second rehearsal runthrough ❏ Finalise contributions to documentation and presentation. Comments back to pitch manager
Stage Five	❏ Manage third rehearsal (and more if required) ❏ Presentation documentation complete	❏ Third rehearsal runthrough
Stage Six	❏ Manage presentation delivery	❏ Deliver presentation
Stage Seven	❏ Manage team debrief ❏ Contact client for debrief	❏ Contribute to team debrief

❏ **The 'B' team.** Not as good as the 'A' team but still have relevant experience and could deliver the project.

❏ **The 'C' team.** Do not have the relevant experience and would struggle to deliver the project.

Now let's look at several project scenarios and let's assume that you have several offices around the country.

❏ **Scenario 1:** There is a very large hotel project in Bristol. The project is sufficiently large to enable you to bring together an 'A' team from various offices to pitch for the project and then go on to deliver the project.

❏ **Scenario 2:** The hotel project in Bristol is medium sized in value and would not justify the 'A' team. However, you have an office in Bristol and they do have some hotel experience and could deliver the project. Therefore during the enquiry process and the lead up to pitch and tender, the local 'B' team can be mentored by your 'A' team. The 'A' team do not appear in front of the client. The 'B' will pitch for the work and if successful go on to deliver the project. This way the client is not let down and gets the team he has selected.

❏ **Scenario 3:** There is a small hotel project in Norwich. You do have a local office but there is no local hotel experience and the project is too small to pitch for it with the 'A' or 'B' teams. The local 'C' team are tempted to pitch for the project. They know that they would pre-qualify using the firm's national experience and CVs of the 'A' team. However, the 'C' team should not try to pitch for work. They run the risk of failing to deliver and damaging the firm's brand and credibility for future work in the area and sector. This 'C' team could well be an 'A' team or 'B' team for another project within a different sector and should pursue those projects.

The audience

The objective in winning a pitch is to win over each and every member of the audience, especially the decision makers.

The pitch manager must find out who will be attending the pitch and then find out each person's role. Are they decision makers or influencers? What are the key things they are looking for individually? Possible points could be as follows:

❏ **The chairman, MD** or a lead client person will want the project to be a success. A failure will reflect badly on him. He wants to be able to trust the key person on your side. Will you be there when there are problems to sort out?

❏ **The client's project manager** will want to make sure you are capable of delivering what is required, on time and to budget. He will want to make sure it will be an easy team to work with. He does not want conflict. Also, he will want to be satisfied that there are adequate resources available and dedicated to the project with the appropriate skill sets.

❏ **The finance director** will want value for money or even a cheaper solution. He will want cost certainty and will want to make sure the project is delivered on or below budget.

❑ **The end users** will want a team that can deliver their wish list, which might be at conflict with the allowable budget. They want to see pitch team members who understand their needs, know the latest techniques and methods and speak their 'language'

❑ **External advisors** might have been employed to develop the brief and help the client through the procurement process. They might have their preferred ways of project implementation. They might have their favourites among those pitching. They will want to be seen to earn their fee so they might be the ones who will look to catch you out and ask the difficult questions.

The content of the pitch

The structure of the pitch is extremely important to tackle all the key issues and perhaps also show off the added benefits your team can bring to the project. You need to be able to differentiate.

The fundamental structure, (see **Table 5.2**), should be as follows:

❑ An introduction, when you **tell them what you are going to tell them**. This will be a brief outline of the content and who will be presenting each section. The introduction will confirm that you have covered all the points that the potential clients have asked for. A structured outline at this stage will give the audience comfort that you are well organised and shows that you intend to stick to your allotted time slot. The introduction is very important, and the audience will decide very quickly if the pitch will be worth listening to. You want to grab their attention, outline the presentation and establish credibility.

❑ **Tell them**. This is the delivery of the presentation and will follow a logical format. During the presentation you will need to address the key issues. You will need to demonstrate the benefits of your proposal and demonstrate your capability and reliability. Your language should suit your client's business and be persuasive, enthusiastic and polished.

❑ **Tell them what you have told them**. At the end of the presentation, summarise the key points and bring home any differentiating factors.

Table 5.2   **Basic structure for a presentation**

Basic structure for a presentation	
Tell them what you are going to tell them	This is the introduction and should include the following: ❑ An introduction of the presentation team ❑ The purpose of the presentation and brief background leading up to the present situation ❑ Confirm if there will be handouts before, during or after the presentation ❑ An overview of the content and structure of the presentation ❑ Give an indication of time (to deliver the presentation, time allocated for questions and follow up discussion)
Tell them	This is the main presentation and should: ❑ Follow a logical format. This might be set out by the client ❑ Stick to the key issues ❑ Demonstrate the features and client benefits
Tell them what you have told them	This is a summary and should: ❑ Clearly recap the key issues ❑ Emphasise the features and benefits for the client ❑ Bring out any differentiating factors ❑ Conclude by asking for the first question

❑ Conclude by asking for the first question. Resist allowing questions during the presentation because you then do not have control of the agenda and time.

Format of delivery and use of technology

Clients may ask you what facilities you would like available? Obviously you will need to decide if you want to bring your own equipment or rely on the client. If there is time to set up and you have good equipment then always go for your own; otherwise you run the risk of something going wrong and not being able to use the controls.

Some clients will ask you to forward your presentation electronically prior to the meeting, so that everything is preloaded and ready to go.

You may like to have a flip chart available just in case there is a need to explain something by diagram during the question and answer session.

Now the big question is, do you use PowerPoint? PowerPoint presentations can be great or they can be awful. If you are going to use PowerPoint then

❑ do not fill the screen with text;

❑ do not use the screen as a script;

❑ if you are going to use bullet points make them short and limit yourself to three or perhaps five per screen (odd numbers always seem to work better).

❑ use the power of illustrations or diagrams and fill the screen with an eye-catching image to make a point, with perhaps one key word to reinforce it.

When you deliver the presentation you need to do the following:

❑ Be enthusiastic.

❑ Make it interesting with a good beginning, middle and end.

❑ Be clear and to the point.

❑ Have a compelling ending.

❑ Show them you care and made an effort to understand their needs.

❑ Make specific points looking at the appropriate person in the audience whom it would concern most. Even use their name such as in the following:

> *John, as the financial director, you will appreciate the savings our proposal will create . . . .*

A good presentation will be a mixture of the following:

❑ Good visuals to endorse the point being made

❑ Use of facts and figures or case studies to demonstrate experience and capability

❑ Appropriate persuasive language

Research has shown that people remember if they see and hear information at the same time. So don't just rely on visuals or just an oral presentation. Use also the power of three to make a point and the power of repetition. Tony Blair famously used

Education, education, education.

Try using the format

Firstly . . . .   Secondly . . . .   and Finally . . . .

Also use your body language, including hand movements, to emphasise points. Obviously, do not be over-dramatic but be appropriate to the presentation, the audience and the venue.

The 'ring master'

During the presentation, you will need one person in charge of the timing. This person will have created a running order and time sheet for the whole presentation. There is a great tendency to over-run on individual slots and if you are not careful you will not have the time to deliver all the points properly.

The ring master will be sitting to one side and will be timing each slot and will have several agreed discrete hand signals such as the following to advise the speaker:

❑ Speak louder.

❑ Speak slower.

❑ Speak faster.

❑ Change the pace of your voice; you are putting them to sleep.

❑ You have 1 minute left.

❑ Finish now.

And when the ringmaster is doing his piece, there needs to be a deputy ring master to help him.

Questions you don't want to be asked

Many people who are about to deliver a pitch feel uncomfortable. They don't want to make a mess of the presentation or to be caught out not having the appropriate answer to questions.

During the rehearsals I always ask my team one by one, *What is the question you don't want to be asked?*

We then spend whatever it takes to find how best to answer that question. We would go over all the various answers and follow-up questions until each person in the pitch team is happy that they haven't got a question waiting to catch them out.

Doing this work helps to boost the confidence of the team and they almost look forward to that question being asked.

Some professionals believe it is a good idea to raise a potential negative factor before the client does. They believe that bringing it out shows you have considered the point and have a good answer to resolving it. My view is, don't raise it unless you are asked, the following being the reasons:

❑ The client might not have seen the point as a negative one. Why bring it up if the client does not think of it as a negative?

❑ If he does raise the point and you have a good answer prepared showing how you have considered the issue, then he is likely to be impressed. He will probably see your answer as being off the cuff and not realise that you have spent hours crafting a response.

❑ If you bring up a negative point for discussion, that might be the point the client remembers and might overshadow all the good positive points.

❑ If you bring up the negative point, then you are not sure where the conversation will go and how negative the point might become.

Handling objections

While preparing for the pitch, the team needs to be talked through the basics of handling objections or negative comments, which may come up during the presentation.

If an objection, criticism or negative point comes up in the presentation, do not rush to be defensive or give off the cuff answers until you have explored the issues behind the client's comments. By determining the real issues you may be able to offset the negative with positive factors. For example, a potential client might ask you the size of your firm and after replying he may say you are either too big or too small for him. Now you need to establish why he has this view. It could be that

❑ if you are a big firm, he perceives you will be too expensive;

❑ if you are too small, you are too inexperienced or lack the resource to tackle his project.

Having established these reasons, you could reply to the first point that the size of your team enables you to put the appropriate level of person onto each element of the project. You have technicians who do the computer draughting and you have experienced job runners to manage the projects. Because the team allocated to the client will be bespoke to the work, the overall fee is very competitive. A small firm might have senior people doing simple tasks, or alternatively not enough senior people doing the important elements of the project.

Regarding the second point, you may answer that you actually specialise within the potential clients sector unlike some of the bigger firms that might cover a whole range of sectors. You may go on to say that you have very experienced people within the client's sector and could illustrate that with CVs and experience sheets.

Therefore the key to handling objections is to get behind the negative comment. You may need to find out what the real issues are and demonstrate that it is not a problem when appointing your firm. Convert the negative into a positive. Also bring out the benefits of appointing your firm whenever you can.

The venue

If you are presenting at the client's venue then you need to send someone to visit the actual room. Also ask the client what the layout will be. If it's in the Boardroom then the layout will be as it is when you do your revisit. When inspecting the venue find out the following:

❑ How much space is there at the front for the presenting team?

❑ What equipment is being made available?

❑ Is there sufficient blackout to the windows so they can see the screen?

❑ Is there a screen or will you be projecting onto a wall? Do you need to bring a screen?

❑ Is there enough room for your team to sit at the front or to one side?

❑ Where are the power supplies? Will you need to bring power leads?

❑ How are the acoustics? Will you be competing with outside noise?

❑ Where are the light switches so you can dim lights if need be?

Rehearsals

This is perhaps one of the most important aspects when preparing the winning pitch.

Team members will find numerous reasons why they do not need to rehearse. They are too busy, they know what they are going to say, etc. The winning teams will rehearse until they are pitch perfect. As a guide I would recommend that you have three rehearsals in the timetable.

The first rehearsal is really a rough runthrough of content, agreeing who will say what and in which order, and to agree on the time allotted for each speaker.

The second rehearsal is the first runthrough. Do it as realistically as possible with chairs arranged at one side, and practice the handovers. At this rehearsal you might go through each person's pitch several times. Also invite an outsider to the rehearsal. Someone whose opinion you respect and who has been briefed on the purpose of the pitch. At the end of the pitch they will need to give constructive comment to the team and cover points like the following:

❑ Delivery – the voice, maintaining interest, did they smile?

❑ The content – was it clear, logical and did it make the desired point or impact?

❑ Timings

❑ The use of technology – was it appropriate, was it professional?

❑ Was it magical?

The third rehearsal, and more if required, will polish the presentation so it's slick and professional and a joy to listen to.

Using persuasive language

The use of appropriate language can greatly assist you in persuading the potential client to place work with you. A very powerful tool is the use of the presupposition. Presupposition is the use of words that give the impression that an idea or experience is presumed without actually saying so. For example,

Have you now started to win work?

This presupposes that you have been losing work.

The following are examples of some phrases that you could introduce in your pitch.

This is one of the cost-saving elements of our proposal.

This assumes that there is more than one cost-saving element.

Another feature, that will deliver distinct benefits, is . . . .

This assumes that there are other features giving distinct benefits.

Looking at your brief I am sure this will be one of the advantages that will really interest you.

This assumes that there are more advantages.

In addition, you can strengthen a point you are making by adding a question on the end. Examples:

You wouldn't want to decide without a fully costed proposal, would you?
This seems the appropriate course of action, doesn't it?
The stakeholders will want to know what is in it for them, won't they?
Your operations director would prefer this layout, wouldn't he?
I can just see the employees liking this restaurant design, can't you?

To maximise the effect, use a downward voice inflection when you say the question. It is generally perceived that a downward inflection at the end of a sentence is a command, whereas an upward inflection is seen as a question and a level voice tends to convey a statement. If you try these techniques you will see an improvement in your influencing methods, won't you?

When trying to persuade people use positive language. Don't use words like possibly, probably, hopefully and likely. Use phrases that include positive language such as 'when we,' 'certainly will' and 'will be.'

Feedback

Win or lose, you need feedback. It is best if someone undertakes this outside of the team and someone who has a reasonable amount of skill and authority to tease out the views from the client.

Feedback needs to be

❑ undertaken soon after the results are known;

❑ with the decision makers;

❑ constructive; you should be able to learn from negative feedback;

❑ reported back to the team in a diplomatic manner, but not so diplomatic that the benefit of the feedback is lost.

Most clients will respect your professionalism in asking for feedback. On larger projects try to obtain a face-to-face meeting. Feedback from smaller assignments can be obtained by way of questionnaire or telephone interview.

5.4   The selection process – direct with the client

Professional firms can be selected by negotiation or competition. Formal competition can also lead to partnering arrangements and framework agreements.

The selection of the professional services provider may be through any one of the following methods.

Negotiation

Negotiation is more common when

❑ the professional services firm is well known to the client, perhaps through working together on previous commissions or through recommendation;

❑ the assignment is small where competition might not be appropriate;

❑ the professional services firm has a definite unique selling proposition, be it vast experience or rare experience within a niche sector;

❑ it is a seller's market.

Negotiation over the services, the terms of appointment and fees can be agreed on using standard schedules of services.

Pitching for the project or through formal interview

Here the client may do the following:

❑ Invite several firms to pitch or present their suitability for the project. The content may be stipulated by the client or may be left to the presenting team to decide.

❑ Invite several firms to attend an interview in front of a selection panel.

❑ Ask the professional team to provide information prior to the formal interview.

❑ After the presentation, the preferred firm may proceed to negotiate the services to be offered and the fees.

Qualification-based selection

The client will have a selection criteria based upon

❑ the quality of previous work;

❑ the qualifications and experience of the firm and the proposed team members;

❑ the suitability of the firm to the project, client's approach and working methods.

For these projects, the client may well advertise the opportunity or approach firms directly to provide the relevant information. The shortlisted firms may then attend an interview or presentation. The preferred firm will then have to negotiate the fees and finalise the scope of service.

Selection based on quality and price

The client, or his advisors, will have set out selection criteria, which include elements on quality and price. The firms submit proposals, often structured to answer a series of questions. The selection panel will evaluate the proposals and will usually have a form of agreed weightings to give a final score. The professional firms that score more than an agreed threshold will be invited to attend an interview for the final selection process.

Selection based on fee (without design)

Here the professional firms are selected individually or as a team based on their fee quote. Selection revolves around the competitiveness of fee.

Fee competition is more common where

❑ public money is being used;

❑ larger, complex projects are being commissioned;

❑ there is a large choice of suitable professional firms;

❑ there is a buyers market.

In this process all the firms submitting quotes will need to have an adequate brief and the services required properly defined.

A variation of this is when a two-envelope system is used. The technical, qualification envelope is opened first and evaluated. Then the second envelope containing the fee bid is opened and there may be a weighting factored into the fee, depending on how good the firm did in the first envelope.

Selection based on a design proposal (with or without fee bid)

This is more suitable

❑ for larger projects;

❑ for a shortlist of up to three teams;

❑ where design is a key feature of the project (architectural or engineering).

Clients may pay a fee to the teams as a contribution to their investment.

A variation of this is a more open design ideas competition, which would be more suitable for

❑ important projects where design is a critical factor;

❑ attracting international interest;

❑ attracting new firms who have not built up a track record.

5.5 Selection process – through and with a contractor

There is a growing proportion of turnkey projects where professional firms are bundled in with the contractor. This could be design-and-build and its many variations such as private finance initiative (PFI), public–private partnership (PPP), etc.

Two-stage tendering

Two-stage tendering is becoming popular, especially to obtain early contractor involvement.

❑ **Stage one.** The client provides the outline information to those tendering. The contractors will price the profit, overheads and preliminaries. Also, they may be asked how they will execute the project and will need to identify their professional team and indicate their team's fee.

❑ **Stage two.** This is often referred to as the pre-construction phase. Here the project team will finalise the design and method of construction and obtain prices from sub-contractors. This enables them to obtain a fixed price lump sum contract.

The professional services firm will modify how it sells its services to suit the procurement method and the selection team. Within larger firms there will be professionals who are more suited to working with the contractor in a design-and-build situation and others more suited to work directly for the client in brief development and a more traditional approach to delivery of the project. The smaller professional firms will not have that luxury.

5.6   The trend for competitive proposals

Clients will usually want to obtain proposals from several professional firms. Once a relationship has been established there may be opportunities to submit proposals as a preferred supplier and thereby not be in competition. But until that relationship is achieved competition is normal.

Clients need to show the shareholders or taxpayers that the selection has been done through a fair process, especially if public money is involved.

Clients want

❑ value for money;

❑ quality service;

❑ a hassle-free project that delivers what was promised on time and within budget.

Competition does not necessarily mean a cheap service. Hopefully, the client will have undertaken some research and come up with a short list. The most common way to arrive at a short list is to have a pre-qualification stage.

5.7   Expressions of interest

Some clients, especially within the public sector, will advertise the project and call for expressions of interest from interested parties. These official announcements will usually ask interested parties to make their interest known. Usually, there will be a requirement to provide some basic information. Frequently, there will be a time limit and you need to get your information submitted to the client in good time.

If you have been tracking the opportunity, you will already be aware of the project and will have made the client aware of your interest. Alternatively, if you just become aware of the project then you need to contact the client and try to be put on the short list for consideration.

Some clients will be pleased to meet with you but many will just ask you to write in, in which case, you need good documentation to grab the client's attention so that he gives you an opportunity to pitch for the work.

If you don't know the client and he is not aware of your reputation then you really do need to consider the following:

❑ What is the real likelihood of your winning the assignment?

❑ Do you know enough about the client to know that he will be a good client?

❑ How many others are chasing the work?

You need to be ruthless in selecting the projects you want to chase. Better to chase fewer and be more focused on winning than chasing too many and not having the resources to do it well.

5.8   Pre-qualifications

Most projects that are being tendered will go through a pre-qualification stage. During this stage the client, or his advisors, are looking to establish the following:

❑ **Are you financially stable?** Often 3 years of company accounts are requested to show you are making a profit and are not growing too fast. Some advisors will impose criteria that the project value should not exceed a certain proportion of your turnover (say a third). There might be questions asking about your bankers, etc.

❑ **Do you have sufficient resources?** There is usually a schedule to be completed requesting a breakdown of your staffing figures, their status and qualifications. Some clients, especially in the public sector, may ask for an ethnic breakdown of your staff to show your recruiting policy. There might be a requirement to supply CVs of key people at this stage.

❑ **Do you have relevant experience?** The pre-qualification will often ask for details (job, client, value, service offered, date) of projects of the same type as being tendered for. There may be a need to supply client reference details. Also, details of any awards and published articles about your work might appear here.

❑ **What technology do you have and how do you apply it?** There will probably be questions regarding your IT, do you have extranets, what software do you use, etc.

❑ **Office location.** Sometimes, there is a requirement to have a local office to the project.

The client will probably want to have a manageable short list that has sufficient numbers to give choice and competition, but not too many so that the applicants do not have a reasonable chance of winning the contract.

It is important to obtain from the client or his advisors how many they hope to short list. Then depending on how much effort is required and your view on success rates, you need to decide if it's worth the effort or are there other opportunities that are more advantageous?

When completing the pre-qualification you need to do as follows:

❑ **Be compliant.** Stick to any word count limits, required formats and answer the question, not the question you think they are asking.

❑ **Do a bespoke submission.** Where the question asks for specific information rewrite any standard text you may have to make it applicable. Clients do not like to be fed general answers which do not totally address the question in its entirety.

❑ **Differentiate.** Show off if you are able to. Sell how you are going to bring added value to the client. How are you different from all the others who may be applying?

When you have finished the pre-qualification have a colleague read through with a critical eye. First impressions count so make sure you submit a quality submission. The document should be professionally presented and formatted; use good quality paper and binding.

5.9   Tendering

Some clients might dispense with the formal pre-qualification stage. They may already be aware of you from previous submissions and presentations. The tender may therefore contain some, if not all, of the pre-qualification information.

The big difference with the tender will be that it will be very specific and contractual. The tender documentation will probably contain the following:

❏ A brief.

❏ Scope of services.

❏ Draft contract of appointment with back-up documentation (sample warranties, etc.). These contracts might be generic and therefore submitted as typical contracts to be entered into.

❏ Programme showing the tender process, anticipated appointment date and milestone events leading up to the completion of the project.

You will need to devote quality time to this documentation to make sure you don't contract to do something which you will not be covered by your professional indemnity (PI) insurance or be too onerous. It would be advantageous to submit extracts that are causing you concern to your PI insurers and possibly your legal advisors for comment.

Your tender documentation should

❏ answer all the questions accurately. Do not use standard text without checking whether it is appropriate;

❏ list any queries that are still outstanding and the assumptions you have made within your submission;

❏ point out any points you are not happy with or which requires future discussion during the evaluation stage;

❑ be very specific as to what service you are providing; also mention what additional services you could offer that are not part of the tender at this stage.

Remember that the main objective at this stage is to get to the next stage of discussion with the client. So don't include more services if you don't need to at this stage. Offer your minimum service for the project for your minimum price. It's easier to offer more at a later stage.

You will need to demonstrate within the tender that you totally understand the following:

❑ **The client's needs.** This may not be totally evident from the tender documents. A smart organisation will have identified the project opportunity and may have built a rapport over many months or years. During this time the key issues may have been discovered. If there still is some uncertainty then try to meet up with the client and explore the needs. There is a requirement to be careful here because the client may, if the questions are asked during the tender period, be under an obligation to tell all the others tendering the same answers.

❑ **The client's market or environment.** What issues influence the way the client goes about his work?

❑ **Where the market is heading.** This is where some clients take comfort if you already work within their sector or work with similar clients. They would like you to help them create a better solution rather than them educating you.

Understanding how clients evaluate proposals and tenders

Many clients will have some sort of evaluation criteria and some, especially in the public sector, will declare how they will score the proposals. Just like answering an exam question you need to address each area that is going to attract marks.

The person managing the tender will need to breakdown the elements of the proposal and make sure each area is answered as fully as possible and at the same time making the answer as bespoke as possible.

Some clients will already have a preconception of the various firms tendering for the work. If all the previous stages have been carried out and you have been building up relationships then, hopefully, the client will be in a positive frame of mind when evaluating your bid.

Some clients will skim read the proposals quickly to get an overview of the submission before looking at it in detail. Therefore, it is important to make a good first impression by

❑ using an easy to read format, using headings and sub headings to break down large areas of text;

❑ using diagrams to reinforce a statement or idea;

❑ using images where appropriate to demonstrate experience and skill on similar projects and assignments;

❑ using the client's structure; imagine that the client will have a score sheet, which will be based on the questions he has asked you to address. Therefore use his numbering and answer his questions specifically.

On larger projects, the submission will be in sections and the client will be dividing up, or copying certain sections, to pass on to others for detailed examinations. The important point here is that you need to make sure each question is answered in its entirety and stands alone. For example, if one question asks how you use technology to create efficiencies and another asks you what IT systems you have, do not assume that the same person will be marking both questions. Not all the markers will have all your submission but perhaps just their section to mark. Therefore if you need to repeat some information, do so for the sake of creating a full stand-alone answer.

When evaluating your submission the client may also be considering the following questions:

❑ Do you really understand their needs?

❑ Are you just repeating what the client has said or are you truly committed to a method of working or approach?

❑ Are you offering a true value for money proposal?

❑ Will he receive the level of service and commitment he wants?

❑ Has the project been thought through and do you have a full understanding of it?

❑ Are the appropriate resources available?

❑ Will he be happy working with you and your team? Do you come over as a business-like professional or someone who has rushed to get a proposal together?

5.10   Using CVs

Within the tender there will often be a need to describe the proposed team to the client. Incorporating CVs usually achieves this.

Throughout all the previous stages we have seen that people buy people. The building industry is a people industry so the CVs and resumes are key selling elements within this stage. A resume of an individual may occur in the main body of the proposal; the full CV will often be required and may be attached in an appendix. The client will take a keen interest in the team that is being proposed and with whom he may have to work for several years.

The CV, as a key selling component, will need to be structured and presented to meet the requirements of the project. Therefore when submitting CVs you need to do the following:

❑ Rewrite each CV, bringing out all relevant information specific to the project being tendered.

❑ Omit information which dilutes the suitability of the person.

❑ Try to make each CV the appropriate length to suit the project. Try, if possible, to have the CVs of all people to be of similar length.

❑ Include a structure of the team showing how they will work on the assignment. This structure incorporates the client's team and who relates to whom and key points of contact, or single points of contact, etc. Do not confuse the client on how your business is structured; he wants to know how it's going to work for him.

On large assignments the individual's qualifications and experience will be fully evaluated and will be part of the scoring system.

Especially critical will be the project manager or team leader. On complex assignments, where the client is looking at the professional firm's ability to analyse the client's needs and understand and cope with complex situations, the team's CVs could win as much as 60% of the technical marks. Within such projects, it is important to indicate the strength of the proposed team early in the proposal document and introduce a matrix of team experience.

CV maintenance

Someone within the professional firm should be responsible for the collation and maintenance of CVs. When new people join the firm, their CVs should be created on the basis of a template that can be adapted for bespoke submissions.

The base CV will never appear in any proposal documents but will be used to extract relevant information for particular assignments. Ideally, about three quarters of the base CV will be usable in the final proposal CV, thereby eliminating the need for major rewriting.

The base CV will contain

❑ personal data;

❑ key qualifications;

❑ professional status;

❑ career summary to date;

❑ experience record (role undertaken, building type, sector areas, client type);

❑ other information such as languages, IT skills, published articles, etc.

Editing

For each project the base CV will be used to create a bespoke CV for the specific project being pursued. You will need to bring out the following:

❑ Projects that have been worked on which have similarities to the one being pursued;

❏ Positive features such as similar role, location, environment, projects worked on in the past with other team members being put forward;

❏ The best possible match to the specific project requirements.

If other firms are joining up to present a team bid, then all the CVs need to be rewritten to make them the same in structure and consistency.

The presentation of the CVs is also important. Include photos, if possible, of the proposed professional and make sure that the photographs are similar to the same style and backgrounds. Do not use photographs taken in photo kiosks!

If the same project is referenced in several CVs make sure the client and project name are the same and not variations on a theme.

5.11   Monitoring progress of the tender or proposal

Once you have submitted the tender or proposal don't just sit back and wait. The bid manager should be always at hand to answer queries and requests for further information.

Having built up your contacts in the various departments within the client's business you may be able to obtain some feedback on the evaluation process. By actively monitoring the situation you might be able to detect preferences and attitudes developing. In such circumstances you need to be at hand to respond to any negative views and to tackle any positive views developing within your competitor's camp.

5.12   Post-tender interview

After the tender evaluation has been completed, the preferred bidder or the best two or three may be invited to attend a post-tender interview.

The client or his advisors will probably set out the structure for the interview, which may follow the lines of

❏ an initial presentation by the professional team addressing key points as outlined by the client;

❑ a question and answer session;

❑ a summing up by the professional team.

The process involved in preparing for the post-tender interview will follow the lines of preparing for a pitch but will be more specific. The client may request that certain people present and even specify how much time is given to each major topic. Where possible, try the following:

❑ To keep each element of the presentation short and to the point.

❑ Always focus on the needs of the client and how your team will deliver those needs.

❑ Don't overload the presentation with too much detail. Just use as much detail to make the point in the time available and in easily digestible time slots.

❑ Use diagrams and other visuals where appropriate.

❑ Always be positive and enthusiastic.

❑ Always answer the questions being asked, don't skirt around issues.

5.13 Negotiation

It is not unusual for the professional team to enter some form of negotiation in respect of their fees and scope of services. Negotiation may occur at various stages during the pre-appointment stage.

Some clients will not go through a competitive tender process but prefer to negotiate with their preferred professional services provider. Other clients, however, will involve the professional in a long rigorous selection procedure and they, once a preferred professional services provider has been selected, proceed to negotiate the final appointment.

Negotiation may involve the following:

❑ The contractual arrangement. There may be a requirement to go through the appointment contract clause by clause.

❑ The scope of services.

❑ The fee.

Establish your position

Negotiating with your potential client might be a quick exercise but, on large projects, could demand a large amount of your time. You will need to establish your current status. Some clients will not be forthcoming with this, preferring to negotiate with you but not weakening their position by telling you that you are the only firm being considered.

Try to find out if the client is negotiating with others, with a view to securing the most for the lowest price.

Negotiating approach

You have probably invested a lot of time and effort in reaching the negotiating stage; therefore, you are focused on securing the project on the most favourable terms to your firm. However, you need to be aware of the client's position and see the negotiation phase as a process to reach an agreement that suits both parties. It should not be an adversarial process. To assist you in the process, you need to spend time re-evaluating your proposal or tender documentation and revisit the client's brief or tender. It is advisable to do the following:

❑ Determine, before entering the negotiating process, your fallback position on key areas of your offering. Identify what scope there is for manoeuvre. You do not want to be caught out having to compromise or concede a position without having considered the situation in advance.

❑ Look for areas where you can offer additional benefits to the client without compromising your position. You therefore need to fully understand the client's needs and be seen to be reasonable and willing to view issues from the client's perspective. You need to make it appear that any concessions on your part involve high costs to you and high value to the client. Equally, any concessions offered by the client may involve low costs to the client and low value to you.

❑ Agree on the technical aspects of your offering with the client before you embark on negotiating the price. You need to finalise, as far as possible, the scope of services, deliverables, programme, project administration and professional inputs.

❑ While negotiating, use your listening skills to look for signs of possible areas of compromise.

❑ Maximise time in between the formal negotiating sessions. You may find that compromises are arrived at more easily in informal environments.

❑ Once you have reached a deal, make sure all the points are properly recorded and agreed to by the parties involved. Do not be embarrassed to clarify all the points discussed and to clear up any ambiguities. You may decide to resubmit the whole proposal document incorporating all the points agreed on, or just agree by correspondence on the points that have been modified since the original proposal. Be certain to obtain confirmation about fees, drawdown schedule, payment terms, etc. Better to know at this stage than come across problems later.

Bargaining skills

To be a successful negotiator, you will need to be able to trade with your potential client so that both parties can achieve their respective goals.

You will need to have prepared for the negotiation and thought through any concessions. Don't rush to give away a position without thinking it through. What might seem like a good idea may prove costly to you in the long term.

When bargaining, be careful to give away just the minimum amount that the client wants. You will know that the client may start his bargaining position, knowing he will need to give way on his position, so his opening requirements may well be inflated.

Be careful not to concede too quickly or too generously; otherwise you may be signalling to the client that you are too keen to secure the deal and thereby weaken your bargaining position. As the bargaining continues, make sure your concessions get smaller and smaller, signalling that you have very little, if anything else, to concede.

When bargaining, be careful not to have used rounded numbers either in your original proposal or during the negotiation. Rounded numbers indicate just that, they have been rounded, usually upwards, and so are a prime target for your client to attack. Try to use figures that appear to have been arrived at through some evaluation, signalling that there is little room for manoeuvre.

When bargaining, trade items that are inexpensive to you but might be seen as expensive to the client. During the process you need to emphasise the value of your compromises and devalue, politely, the client's compromises.

While negotiating, don't lose sight of the big picture. It may seem reasonable to give way on a series of small issues but never finalise until all things are agreed. Otherwise the client may cherry-pick the points he wants you to agree on but never seems to give way on points you want movement on.

During the negotiation, the client might make unreasonable requests or demands. Don't reject the client outright, but instead reply with your own request for a concession in return. Equally, if you propose a compromise and the client says no, then ask him to put forward an alternative. This helps to keep the negotiation moving along.

Summarise the situation during negotiation

Make a point to summarise the position as you see it at various stages during the negotiation process. This has the following benefits:

❑ Clarifying the situation, especially if you have been through complex issues and offers and counter-offers. This allows both parties to check that they have the same understanding of the current position.

❑ Keeps your focus on the key points and makes sure minor issues do not hijack the negotiation.

❑ Allows you time to consider your position and signals to the other side that you have been listening.

❑ Keeps you in control of the negotiation and lets you set the pace.

❑ Helps both parties to focus on key stumbling points.

Don't get stuck over positions

Negotiation is when both parties take up and then give up on a series of positions. Sometimes negotiations get stuck in a position that the negotiators feel obliged to maintain and then feel that they cannot give way because it will be seen as a sign of weakness and loss of face.

In such situations, it is better for the party to devote more time to discuss the underlying reasons or concerns that the other party may have. It is better to single out these underlying concerns and resolve them rather than just meeting people 'half way' in the hope of securing the deals. The project you secure may take several years to deliver, so it is vital to iron out and resolve legitimate concerns at the outset.

Arguing over a position within a round of negotiation can

❏ be unproductive and risk a breakdown in negotiating;

❏ jeopardise an ongoing relationship;

❏ end up with an unsatisfactory compromise.

So when the negotiations stall, due to a position taking, try the following:

❏ Focus on the problem, not people. Try to get both sides to work together to sort out the problem.

❏ Generate a variety of options for consideration if possible.

❏ Try to base the decision of moving forward on some objective criteria, for example, on how the service would satisfy the needs and provide additional benefits.

In the discussions, tell the other party the reasons for a problem rather than coming out with the objection first. For example, instead of saying

I can't possibly agree to that contractual term within our contract

you could say,

> *As you know we carry professional indemnity insurance cover, which is also one of your requirements. We always pass contracts over to our insurance brokers just to make sure there is nothing we have in our contract which negates our cover. Our advisors say that the phrase you are proposing is not acceptable to them and would not be acceptable to any other professional service provider's insurers. They are proposing an alternative phrase, which I believe covers the point you are making and still gives us the insurance cover you also require. Therefore you are in a win–win situation. Is that acceptable to you?*

During negotiations always try to encourage all the parties to look forward to resolving any obstacles rather than being entrenched or even going backwards. It may be useful to take a break and do an overview where you can emphasise the success so far and encourage all parties to come up with a solution. The solution does not necessarily mean one party winning and the other party losing. There could be an imaginative alternative option for mutual gain.

Move to closing the deal

When you are aware of the appropriate buying signals, then as soon as possible, move to close the negotiation. If the potential client is saying 'yes' in words or body language, you may move him out of the buying mode by continuing to try to sell. If you pass the buying mode, you may raise doubts or put forward something else for him to consider and jeopardise the sale. Indications of the client being in the buying mode are as follows:

❑ Being very receptive to your pitch, asking relevant questions and appearing satisfied with the answer;

❑ Saying things like 'yes that's just what we need';

❑ Being very attentive in their listening and good body language such as nodding and leaning forward, making notes, etc.

Once you have seen these positive signals then ask for the business. Many professional services consultants fear asking the question in case they get a 'no'. If you have moved through all the stages in this book and adequately established the client's needs and in turn put in a proposal that satisfies those needs with sufficient benefits and value for money, then you need to know why the client is saying 'no' sooner than later. Could it be one of the following?

❑ He hasn't the authority to close the deal then and there.

❑ There are a few points he needs to clarify or seek other people's advice on.

❑ He wants to consider further options or modifications.

❑ He is also considering other proposals.

So if you do get a 'no' when you ask for the business, try to establish the reasons for the 'no' and help the potential client resolve any issues he may still have. Do this as soon as possible and keep the competitors out.

When faced with a 'no' and the potential client tells you what the outstanding issue is, then try to use some more searching questions to establish if you are still in the frame. Ask questions such as:

If we could resolve that issue what then?

The use of 'what then' is a useful phrase, but do not overuse it. Alternatively dispense with the 'what then' but ask your question and then raise your voice for the last few words. The raising of the tone indicates a question and will give you a reply.

Negotiating traps

You need to be aware of potential problems that occur during negotiation. These include the following:

❑ Are you dealing with the right person? You may spend many hours negotiating and making concessions only to find out that the other person needs to refer to his boss. This may involve you in another round of concessions. Therefore, at the outset, establish from the other side if they have the authority to agree on solutions reached.

❑ Pressurising you to agree due to time running out. Don't be unduly pressured to agree or compromise because the other party says he has a deadline to meet. Establish the ground rules at the outset and decide if the pressure is reasonable. If it is not, say so, the client may concede if he knows that a better deal may be available if both parties have sufficient time to consider the key points.

❑ Clients might suggest that you meet them 'half way' on an issue. But be aware that 'half way' might not be good for you and the client might have inflated his opening stance so 'half way' is good for him.

If you fail to win, start positioning for the next opportunity

You may think you have submitted the best proposal or the best price or both and yet, for some reason, you fail to secure the project. If you still believe that the potential client has other projects in the pipeline, and you still want to work with him, then you need to start positioning yourself for the next opportunity.

Therefore you will need to do the following:

❑ Ask for feedback to establish where you could improve your offering next time.

❑ Try to establish why the competitors got the project. Some clients and most public sector clients will give you a score feedback based on their criteria. They will also give you an indication on how you were placed against your competitors.

❑ Tell the potential client that you would like another opportunity to work for him and ask whether they would consider you for other projects.

❑ Find out what other projects there are and put the details into your pipeline and track the opportunity.

❑ Keep tabs on the success or otherwise of the project you lost. Learn from your competitors. The more you learn and feed back into your proposals, the better your proposals will be and your success rate will be improved.

Summary check list

By the end of Stage 5 you will have responded to enquiries by doing the following:

❑ Submitting proposals

❑ Pitching for the work by presenting

❑ Submitting expressions of interest, pre-qualifications and tenders

Within the professional services business you will know that your key asset is the people within your business and team. You will have created a bank of CVs, which can be fine-tuned for each specific opportunity.

If you have been successful you will have probably gone through a negotiation stage to finalise the contract and the scope of services to be provided. You will have developed your

❑ negotiating approach;

❑ bargaining skills;

❑ ability to recognise when to move to close the deal.

Stage 6: Delivering added value and obtaining repeat business

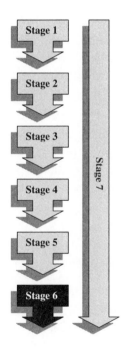

Stage 6: Delivering added value and obtaining repeat business

6.1 Obtaining repeat business

Before considering securing repeat business take some time to thank your client for the project he has just given you. Consider the following:

❑ A thank you letter from your senior partner, MD, chief executive or chairman. The letter has to be sincere, bespoke and not a standard letter. It should not matter how big the project is. If you have accepted it, then you must have considered it worthwhile. So make the client feel important from day one. Make him aware that you are very appreciative of his business.

❑ The thank you letter or a follow-up letter should outline, and remind the client, who in your team is in charge of the project and who are the other team members and what are their responsibilities. Also, advise the client whom he should contact if there are any problems with the service delivery.

❑ Also, in the thank you letter, put in place the idea of regular meetings, at senior level, to review the progress of the project. These review meetings could be at significant milestone events or say every 3 months.

Consider also having some cards printed which can fit into a standard envelope. You can then send personal, handwritten notes to the client to show you are still involved and monitoring the project. Messages can be short but will let the client know that you still consider him important. Have the delivery team prompt you when certain milestone events occur and also when problems arise. Keeping in touch with the client, at senior level, during the project is critical to the smooth running of the relationship and to becoming aware of any repeat business opportunities.

Repeat business is by far the best kind of work to be receiving because of the following:

❑ The client knows you and is obviously happy with your work to be offering you repeat work.

❑ You know the client, his systems and procedures so there is little or no learning curve.

❑ There are no or little marketing costs.

Once you secure work for a client you should be trying to win more work, initially by delivering a good service and secondly by getting to know the client better so that you know in advance, and are prepared for, any up and coming work.

Very often the professional services firm will lose interest with a new client. There will be the initial excitement of winning new work with a new client. The project is handed over to the implementation team who may or may not have been involved in securing the work. The pressure is now on to deliver the project, keep the client happy and make a profit. Sometimes the initial client contact keeps in touch, but, all too often, he has pressure to secure more work and this invariably involves chasing new clients. Sometimes the delivery team don't want the 'salesperson' to continue his involvement with the client. If this is the case, then someone should take over the client relationship interface from the 'salesman', if he is moving on to secure other new work.

The process of winning a new client can be time consuming and costly. Therefore, once secured, it seems sensible to try to secure repeat work. The client team should review the situation regularly and ask themselves the following questions:

❑ Are there any other areas within the client organisation where we could secure additional work?

❑ Are we aware of all the likely opportunities that may arise in the short-to-medium term that could generate opportunities for us to obtain further work from the client?

❑ Is our relationship so good at the moment that the client will give us all additional work he has?

At the same time you need to also ask yourself if, now having worked for the client, is he a profitable source of income and would you be pleased to work on his future projects?

6.2   Strategies for repeat business

To obtain repeat business we need to keep giving the client what he wants and even offering additional services that might provide him with further benefits. You need to continuously maintain and enhance client relationships. There are several ways you can do this (see **Table 6.1**):

❑ **More of the same.** This is where the initial service is continued but with the enthusiasm and freshness as if it was a new client.

❑ **More customising.** Once you get to know the client and how he wants his service delivered, you can tweak your procedures and processes to suit. For example, how does the client want to be kept informed? Headline reporting or detailed reports? Therefore match your reporting procedures to how he wants to receive the information.

❑ **Additional services.** Find out if there are any other opportunities for you to offer additional services. Introduce your client to these extra services, some of which he may not have known you provide.

❑ **Better pricing.** Now that you know your client and how he operates, there is less risk in the service delivery. With repeat work there are no marketing and selling costs, which are usually attached to new clients. Therefore why not let your valued clients share in some of the savings you are making. Offer a more favourable pricing structure to your best clients. Tell them you are doing this. Let them know that they are benefiting from the good business relationship you have with them.

6.3   Preparing a project client plan

Some people will like to manage their client on an ad hoc basis. They believe that they are able to keep their finger on the pulse as to how

Table 6.1   **Strategies for repeat business**

Strategies for repeat business	
Original project	To obtain repeat business the original project needs to be ❑ Well delivered with good client care ❑ On time, on budget and appropriate quality ❑ Meet or exceeded client expectations
More of the same	Build on the experience of the original project by ❑ Same team ❑ Price match ❑ Continue to excel and improve
Customise	Client becomes aware that the service delivery is being specifically tailored to suit his needs by ❑ Service provider modifies services and processes to suit client ❑ Continuous feedback obtained from client and acted upon ❑ Service provider becomes an expert on the client's business
Enhanced relationship	The inter-business relationship improves by ❑ Frequent meetings between service provider and client (not specifically project related) ❑ Getting to know more people within the client organisation ❑ Service provider team becomes more integrated with client's team (share same offices, secondment, workshops etc.). The service provider becomes so knowledgeable about the client's business, and needs, that he becomes irreplaceable ❑ Social events
Additional services	Spot opportunities to offer additional services
Shared benefits	With repeat business there are savings to be made which are shared with the client

the project is going and how the client perceives the service delivery is being delivered. Perhaps with small client organisations this might be acceptable. On large projects, there is a possibility in losing opportunities to capture additional work such as an extension to the current project or missing opportunities for securing new projects.

The professional service provider who likes a process or a methodology to assist in client account management, especially arising from current projects, may wish to develop a project client plan. This is a simple method of recording all the information possible about the people involved within a project and then to plan how that client relationship can be developed.

A typical structure for the project client plan could be as follows:

Part 1: Organisation details
List under subheadings all the information you can about the client (see also capture plans, Section 1.5. If you have already started a capture plan then expand that into a project client plan). Information to include would be basic client contact details, web address, corporate information such as policies, vision, mission and annual reports, etc.

Part 2: Contacts within the client organisation
List all the people you know within the client organisation who relate to your project. How do they interrelate? Who reports to whom? What are their duties and responsibilities? Who knows them in your firm? Also include useful information such as hobbies.

Part 3: Other consultants
Perhaps you have come across new consultants on this project. This would be a good opportunity to expand your network. Are they able to put work your way or introduce you to potential clients? List all the consultants in this section and treat each one as a potential client or a referrer.

List also other consultants whom you know have worked with the client in the past or currently on other projects. If they are competitors, then keep your eye on them so that they don't take the next opportunity from under your feet. Note any comments your client may make about them, good or bad. Make use of this feedback in developing an enhanced service delivery.

If the other consultants (not currently in your team) are not competitors then again expand your networking to include them as well.

Part 4: Services offered and services extension opportunities
Record all the services you are currently providing, all the services you have provided to the client in the past and any additional services you consider the client may want to purchase in the future.

Part 5: Developing issues
Develop a history of issues raised by the client who may help you to improve the current service or make your next offering more attractive to him. The client may reflect, and share with you, how he may have approached the assignment differently with the benefit of hindsight. Alternatively, your client may advise you that he regretted not including additional work in the current assignment. Knowledge is power; listen and take note and act upon your findings. Be thirsty for information and record it. Then when you are preparing your next pitch tap into this reservoir of information.

Part 6: Project SWOT
Prepare a strengths, weaknesses, opportunities, threats (SWOT) analysis for the project. Identify your strengths, weaknesses, opportunities and threats. What action do you need to put in place to maximise your opportunities, build up your weaknesses and eliminate your threats?

Part 7: Benefits and innovation giving added value
Record as you go the benefits derived by your client from your service delivery. Also record any innovation and added value you have brought to the project. You will be able to use this to remind the client that he made the right choice in selecting you, and use the information when pitching for more work for the same client. Also this will be good information for project sheets and case studies. All too often this detail is forgotten or overlooked at the end of the project.

Part 8: Opportunities identified
Record opportunities to do more work, either by project extension or new projects. See how the opportunity develops and be ready for the right moment (switch to buying mode) to pitch for the work. Perhaps you could offer some free advice to get you into the opportunity.

6.4   Total continuous office participation in selling

The smaller professional service firms will not have the resources for a full-time dedicated business development or marketing person. In these smaller businesses, marketing and selling is slotted into the busy schedules of those who are busy doing fee-earning work.

In large professional services firms there will be distinct marketing departments and business development managers or directors. In these situations, the dedicated resources are there but sometimes they do not act together in concert. All too often marketing and business development plans are drawn up and one is not linked to the other.

So it seems that, at best, a firm will have dedicated people designated to marketing and selling but very few will go that extra step and nurture the whole office as business ambassadors. We mention later the need to manage and enhance the client 'touches' with the firm (see Section 6.5); there is benefit for going further and harnessing your entire workforce to be 'salespeople'.

The client or potential client will have a perception of your firm. This perception will be created through

❑ all the service touches with your firm;

❑ external feedback from other professionals and clients;

❑ media coverage and other PR activity.

The professional services firm can influence all of these. A good place to start is to create the appropriate culture and working environment.

Quality of employees and working methods

During boom periods it is sometimes difficult to retain good competent employees. The professional services firm needs to have the following:

❑ Have the appropriate person dealing with the assignment.

❑ There needs to be the appropriate level of supervision and checking of work before it leaves the office.

❑ Have a development plan to train up and promote employees to better positions.

A happy employee will be more enthusiastic about his work and this will be detected by the clients. Therefore the professional services firm will do well to

❑ have projects properly resourced and not to have employees over pressured, which could lead to stress, mistakes and bad relations with the client;

❑ develop employee commitment to the business;

❑ empower employees to make decisions up to their level of competence.

By developing all these areas the professional services team can create an organisation whose employees can be seen to be an ingredient of competitive advantage.

There are various ways that the employees can be made to feel better about the business in which they work. Involving employees more could be one area to address. Keep employees up to date with new projects and new clients. This gives the employees something to talk about when they next meet their clients. Nothing gives a client more confidence in a professional services firm knowing that other similar clients are appointing the same firm. Also by sharing more information with employees it enables them to understand and contribute to the firm's overall performance.

To facilitate this, there could be more opportunities for team meetings to discuss each other's projects, internal workshops to discuss how improvements could be made to service delivery and client satisfaction.

During recessions or periods of low workload the employees need more encouragement. An employee who is worried about job security might not be as focused on service delivery as he might be during better times.

6.5   Managing the service 'touches'

The service 'touches' are any time when the client directly experiences the professional service firm's organisation.

These service 'touches' might be the following:

❑ Correspondence – reports and drawings

❑ Telephone calls (receptionist, PA, service provider), e-mails

❑ Face to face with any one of your employees

The quality and consistency of these 'touches' are important to the success of the relationship. Examples of how the service 'touch' can provide negative impressions could include

❑ a long wait for the telephone to be answered;

❑ poor telephone manner;

❑ mistakes in communication (reports, letters, e-mails) such as spelling and factual errors;

❑ bad-mannered employees;

❑ delay in responding to client requests.

There needs to be checks in place to make sure that these bad experiences do not happen. Also in respect of employees you will need to review the need for

❑ training in client care;

❑ staff motivation;

❑ better checks and procedures.

If not addressed, there could be a gap developing between client expectation and perceived service delivery. The gap in service delivery is shown in **Fig 6.1**. If this gap becomes intolerable, then the client will defect to one of your competitors at the earliest opportunity.

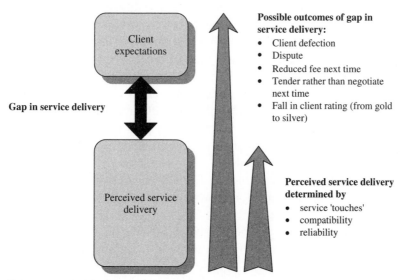

Client expectations determined by
- sales and marketing pitch
- past experience
- referral or reference

Client expectations

Gap in service delivery

Perceived service delivery

Possible outcomes of gap in service delivery:
- Client defection
- Dispute
- Reduced fee next time
- Tender rather than negotiate next time
- Fall in client rating (from gold to silver)

Perceived service delivery determined by
- service 'touches'
- compatibility
- reliability

Figure 6.1   Possible outcomes arising from gap between perceived service delivery and client expectations.

6.6   Client account management

The first key step is to nominate someone in your organisation who will have the responsibility to manage the client account from the point of view of the following:

❑ Keeping the client happy with the service delivery.

❑ Developing a strategy for creating strong relationships with the client team and building a wider network of contacts within the client organisation. This will also include building a knowledge bank relating to key aspects of the client organisation, which will help in preparing to secure additional work.

❑ Finding opportunities to promote the skills within the firm to the appropriate client contacts.

❑ Keeping up a dialogue on the clients' future plans.

❑ Being aware of any competitors trying to make inroads into the client organisation.

Client management is time consuming and therefore there is a need to prioritise the effort towards the most important clients with the most potential to provide future work. We have already discussed gold, silver and bronze categorisation of clients and we can continue this theme when looking at client account management. In addition to the criteria for gold clients mentioned earlier, we can pick out the most important gold clients for enhanced attention. Factors that could influence this choice could include the strategic value of the client and his future workload potential (lifetime value).

There may also be key clients who are extremely valuable to the business. They may not be the most profitable but could be providing a significant amount of business without which the business would have problems.

A healthy relationship needs to be maintained. One of the signs of a relationship that may be failing is the absence of complaints. The client either does not want to raise issues or is not being contacted and being asked about service delivery. If this is left too long, or not investigated, then the relationship might fail. A competitor might be paying more attention to your client than you are!

Therefore the professional services provider needs to be more positive and proactive, such as the following:

❑ Telephone the client regularly rather than wait for the client to call.

❑ Use the telephone more than correspondence.

❑ Suggest service enhancement rather than wait for the client to ask if you can provide additional services.

❑ Speak to the client about future activity rather than rely on the past to get you repeat work.

❑ Uncover problems and sort them out before the client brings them to your attention.

To manage the client account effectively, the professional service provider must put systems and procedures in place to make sure that he becomes aware of any problems or opportunities that arise and to act accordingly. There needs to be regular reporting by internal client account managers on the 'health' of the client relationship.

Managing the process

The professional services firms are usually poor at managing a process for maintaining and managing their relationships with their gold and silver clients. I specifically stipulate gold and silver clients for the following reasons:

❑ Time is limited, so should be focused on the real important opportunities. These would typically be the existing gold clients.

❑ The task is easier with gold clients because, by definition, they like you and are willing to give you repeat work.

❑ You should not ignore your silver clients. These are potentially your future gold clients. Also there is a risk that your established gold clients might be coming to the end of their lifetime value, so you need to replenish your stock of gold clients.

When introducing a process there should be great care to use it first as a pilot scheme or in a limited format. Better to have a lower target and hit it than to aim to do great things and fail. Many professional firms rely on their senior management to maintain client relationships. These might be the original people who secured the client or the professional now handling the client's work.

Each person will have a different amount of available time, or enthusiasm, for true client account management, which is over and above ordinary service delivery. You need to capture existing clients in between projects. You also need to secure work from other parts of the client's business.

Being selective with time expenditure

Before embarking on investing more time on developing existing clients, you need to re-examine your portfolio and double check your gold, silver and bronze ratings, as clients can easily slip down to silver and bronze if not nurtured. The gold clients received your best service and attention and in return they rewarded you with good fees, and repeat work. So, before homing in on the client list you need to revisit their status. You need to do the following:

❑ Check on the amount of work they have been giving you. If they have dropped from the top clients list is it because they had no more work to give out or had your competitors secured it? Do you even know? Had your attention been elsewhere that you missed an opportunity that was snapped up by a competitor?

❑ Check if the work is still profitable. Have you been trapped with rates that you had agreed to many years ago and are now finding it difficult to raise them?

❑ Check if the client is still strategically valuable to you. Does some work, perhaps small amounts, keep you active in a certain sector which could be beneficial in the long term?

❑ Check where the client is in the life cycle of giving out work. Is the client now coming to the end of its building programme or is the client in a dying market and is diminishing in size? You might be able to gain work if the client starts to downsize!

❑ Check if the brand value of the client is still a good marketing asset. A household name on your client list gives other clients, especially in the same sector, confidence in placing work with you.

Having re-evaluated your client list into gold, silver and bronze may highlight your target areas for enhanced account management. If you only have a small amount of gold and silver clients then that is where your attention will be. Continue with your gold clients and try to promote the silver to gold.

If you are in a favourable position, through good business development and good service delivery, and have a long list of gold and silver clients, then you have to manage carefully your limited available time.

It is not being suggested that all the other clients are to receive a poor service delivery. They will still receive an excellent service. You will still deliver your promises and hope to exceed them. You would still carry out all your customer care services. What we are trying to establish here is a group of clients that you can identify as being

❑ critical to your current business;

❑ critical to future growth and survival;

❑ critical to your strategic plan for future development.

This group of clients need an additional investment of your time; they need a proactive input. You will need to home in on this select group and find out what more you could do for them and how the additional activity would be received and valued. Some clients do not want any more 'marketing or relationship building' calls; they just want an excellent service.

So, as in winning new clients, you need to ask all established clients their current needs and future needs. Are their needs changing? Is your once leading edge service now being perceived as a normal service? Are the key client contacts still there? Do you know who will be taking over from your key client contact when they leave through retirement or for whatever reason?

6.7   Client account teams

If your client is a complex one in that you are dealing with numerous divisions, each of which has the capacity to give you work, or you are progressing through various assignments, then your client account management may also be complex. In such a situation, one person may find it difficult to manage all the relationships within the client organisation. Therefore on these occasions there would be a need for a team effort, a client account team.

The client account team can have membership from just your own firm or externally, if you are part of a large service team including external consultants. The key point is to have the right people who are motivated to contribute to client management, for the benefit of the entire group as well as their own sub-team.

Let's assume a totally internal team. The team members could be as follows:

❑ **Key account manager.** This person is in overall charge of the client relationship. He will need to be aware of all relationship activities and make sure they are co-ordinated and fruitful.

❏ **Project team representatives.** A representative from each current project, or identified future project, is part of the team.

❏ **Capture plan and project plan administrator.** This person will make sure that all intelligence about the client is captured (see Section 1.5 for capture plans and Section 6.3 for project plans). They will also chase information from project teams (minutes, reports, file notes, etc.).

The frequency of client account team meetings will depend on the work in hand. There is no need to meet if nothing has happened since the last meeting. To make the meeting worthwhile you need to do the following:

❏ Distribute as much information as possible prior to the meeting. Capture plans and project plans are updated and updates on actions circulated.

❏ Reports are circulated prior to the meeting in respect of any discussion items. The report must include background information, the topic and recommendations. At the meeting the report authors do not read out their reports. This is taken as read by all those attending.

❏ Any other business (AOB) should be restricted and team members should be encouraged to make them main agenda items with back up papers.

All the team members should have read all the paperwork prior to the meeting, leaving key decisions and discussion for the actual meeting. This is more critical in a large firm where team members may be travelling from other offices within the organisation.

A typical agenda could be as follows:

1. Attendance, apologies

2. Previous minutes

3. Actions (updates on actions will have been circulated prior to the meeting. There is no need to cover old ground again)

4. Problems with the client relationship since last meeting

5. Successes with client relationship since last meeting

6. Opportunities arising and any relationships that needs to be created or strengthened to increase success

7. Discussion on reports submitted (already circulated)

8. AOB.

Creating new relationships

The client account team may identify new opportunities for further work, so there is a need to decide the following:

❑ Who are the key people within the client organisation? Do we know them?

❑ Who would be the most appropriate person within your firm to create or strengthen a relationship so as to capture this opportunity? This might not be a current client account team member. If the new prospect involves a new service offering then someone from that service department should be coached into creating a relationship. This may require others in the firm helping out with introductions.

❑ A plan of action and who needs to be involved. Remember the inter-company relationship is stronger if there are many people from both sides that are building relationships rather than relying on one client champion.

Cross-selling

Perhaps one of the biggest benefits of a client account team is the creation of cross-selling opportunities. The team should be motivated and encouraged to identify and facilitate cross-selling opportunities. There is a need to break down the silo approach where clients only get to know one part of the service organisation. Many professionals dislike introducing other departments or offices to their clients. They fear that the current relationship may be damaged by poor performance by others.

The client account team must nip this in the bud if spotted. There is no room for personal empires at the expense of overall team gain.

Cross-team activity

The teams should also be encouraged to interact with other client account teams, especially within the same market sectors. This would encourage

❑ expanding best practice;

❑ comparing sector activity information;

❑ possible opportunities to facilitate clients' meetings to benchmark.

A team member would also feed into a market sector group that would be looking at gaining more work from the sector and strengthening their expertise and skill set.

6.8   Establishing level of client satisfaction

Client satisfaction is a relative measure. It is a measure of how your business 'total service' performs in relation to a range of client needs and requirements. The only way to find out how satisfied the client has been is to ask him. To do this you will need to

❑ ask the right questions;

❑ ask the right people;

❑ adopt the best survey method based on resources available to you.

From these findings you will need to

❑ arrive at the correct conclusions which will tell you how good you were and how you could improve;

❑ draw up an action plan to address the findings;

❑ implement the action plan.

Ask the right questions

The client will have experienced your service over a period of time, perhaps over more than one assignment and having dealt with many members of your team. The client will have established a feel or view of your business based on the outcomes, benefits and results of your service. From your viewpoint, you will want to know the client's opinion or views on your services, people and processes. The following key question needs to be asked to the client:

> *What is important to you and how well have we delivered our service to achieve what you wanted?*

If you asked that question then the reply might not address all the aspects you want feedback on. Therefore, there is a need to structure the questioning to obtain the information you require.

Ask the right people

You will need to select the right people to obtain worthwhile feedback. The aim is to find out the views of those who matter and have a say in your re-appointment. Therefore you will need to select a sufficient number of your client portfolio to obtain a truly representative view of the whole client portfolio.

If resources are limited then select people from

❑ your gold and silver client list;

❑ a spread of sectors (rather than just one sector);

❑ a spread of projects that reflect your portfolio of services.

On the basis of the above, select the level 2 contacts (see Section 2.2) – those individuals who have interfaced with your business the most and who understand the needs and can evaluate the level of service delivery they received.

If more resources become available then widen your selection to include

❑ more of the same – perhaps more from each organisation;

❑ fellow professionals;

❑ bronze clients (those you are hopeful to convert to silver and gold).

The more people who are questioned then the better the quality of your feedback. This way those clients who have extreme views do not sway the overall view. A sample size of 200 would begin to give you a more representative answer. Don't worry if your sample is much less. Some feedback is better than no feedback.

Also within the sample to be surveyed try to obtain a mix that reflects your client organisation. Select from the following:

❑ The large, medium and small client organisations based on the turnover they give you

❑ The large, medium and small client organisations based on their overall turnover (a large client may still be a small client to you, whereas a medium organisation could be your main client because they give you all their work).

❑ Your portfolio to match your sectors. So if education dominates your turnover then your sample numbers from this sector will reflect this.

Survey methods

There are a number of survey methods to obtain feedback. These are as follows:

❑ Self-completion methods. This includes posting a questionnaire, or by e-mail. This is a low-cost method and tends to have low response.

❑ Interview, either by telephone or face to face. Telephone is less costly and should not last more than 15 minutes. Best is to pre-book as you get better results. Face to face is by far the best method as you are likely to get a more thorough feedback and gives an opportunity to probe more to identify reasons behind comments.

Need to introduce the survey

If you make your client aware that you are embarking on a survey then it will improve the response rate and the quality of the feedback. The best way is to send them an introductory letter which will cover the following:

❏ Why you are conducting the survey

❏ Why you would like their particular view and participation (you want to gauge how you met their particular client needs, and you want their help to improve your level of service, so you value their feedback)

❏ How the survey will be conducted (telephone or face to face), offer them a preference, tell them the approximate duration of the survey

❏ What feedback they will get after the survey (overall findings, etc.)

A typical letter covering the above points would be as follows.

Dear Mr Cuthbert,

Re: Client feedback

Wren & Barry Architects are continuously looking at ways to improve client service and project delivery. To help us focus on the areas, which are really important to our clients, we have commissioned an external agency to conduct a feedback survey on our behalf.

We have enjoyed working with you and hope to continue providing you with a service that is tailored to meet your needs. We would therefore be pleased if you would take part in the survey and give us your views.

I have given your details to Client Feedback Surveys Ltd, and David Barnes, one of their directors, will be contacting you in the next few

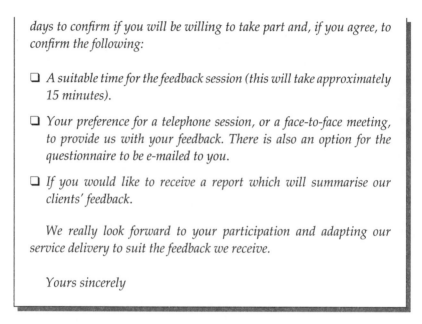

days to confirm if you will be willing to take part and, if you agree, to confirm the following:

❏ *A suitable time for the feedback session (this will take approximately 15 minutes).*

❏ *Your preference for a telephone session, or a face-to-face meeting, to provide us with your feedback. There is also an option for the questionnaire to be e-mailed to you.*

❏ *If you would like to receive a report which will summarise our clients' feedback.*

We really look forward to your participation and adapting our service delivery to suit the feedback we receive.

Yours sincerely

Questionnaire design

The questionnaire should be structured to obtain the maximum beneficial feedback within the time frame available. Also make sure the questions are not ambiguous and are neutral, i.e., you are not pushing for the answer you want. When you have drafted the questionnaire, try it out on colleagues so that you are sure you keep to the tight time frame and have also allowed for opinions to be aired.

Ideally, you want to obtain feedback from each client so that you know how satisfied they were of your service and, equally, how you were rated overall by all the respondents. There is also a need to apply some form of weighting so that the most important needs have the most impact, or weight, on the results.

Introducing weighting factors

To make the most of your clients' feedback there is a need to apply a form of weighting to the answers.

While preparing the questionnaire you will have decided on a range of perceived customer requirements. **Table 6.2** shows 10 requirements

Table 6.2   Establishing levels of importance for elements of service delivery

Importance of elements of service delivery		
Customer requirement	Importance score	Weighting factor (%)
Fee level	7	10.61
Response times	5	7.58
Quality of team	5	7.58
Presentation of design	4	6.06
Contract administration	7	10.61
Being kept informed	8	12.12
Providing added value	9	13.64
Company information	4	6.06
Staff helpfulness	7	10.61
Delivering promises	10	15.15
Total	66	

and the degree of importance according to one respondent (where 10 is very important and 1 is not).

Having recorded the levels of importance attached to each element of service delivery, you need then to ask the client to score your level of service delivery. Against each requirement, record the client's response where 10 are completely satisfied and 1 is not satisfied. **Table 6.3** gives an example of a response.

Having arrived at these findings you can see where you are scoring well or not so well. By comparing how you have performed against the levels of importance the client was attaching to each element of service delivery, you can see where the priorities for improvement are. **Fig 6.2** shows the priorities for improvement for an individual client.

Satisfaction levels are relative

For the client in the example above the overall satisfaction index was 57%. Is that good or bad? Obviously you are able to address the areas you are not so good at, and you are able to prioritise you efforts. However, you need to know how you performed compared to your competitors. If they are all in the 30–40% range then you have nothing

Table 6.3   **Arriving at a satisfaction level for clients**

Calculation of client satisfaction level			
Customer requirement	Satisfaction score	Weighting factor (%)	Weighting score
Fee level	8	10.61	0.85
Response times	8	7.58	0.85
Quality of team	6	7.58	0.46
Presentation of design	4	6.06	0.24
Contract administration	5	10.61	0.53
Being kept informed	5	12.12	0.53
Providing added value	3	13.64	0.41
Company information	3	6.06	0.18
Staff helpfulness	7	10.61	0.74
Delivering promises	6	15.15	0.91
Weighting average			5.70

Satisfaction level for client			57%

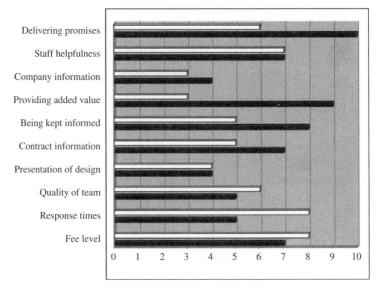

Level of satisfaction received by client

Importance of service delivery perceived by client

Figure 6.2   Comparing importance of elements of service delivery with client satisfaction received.

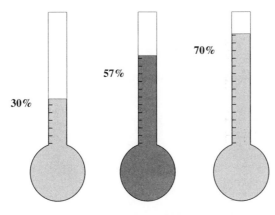

Competitor A   Your satisfaction rating   Competitor B

With a satisfaction rating of 57% you are well placed compared to competitor A,
but need to improve your service compared to competitor B

Figure 6.3   Comparison of client satisfaction levels.

to worry about. But if most of them are in the 60–70% range then you
have severe problems that need addressing (see **Fig 6.3**).

If you have limited resources and time you should do the following:

❑ Tackle, and put right, the service delivery for those clients where
your competitors have a higher satisfaction rating (see third party
survey to establish who they are).

❑ Address the areas that clients say are more important and where
you are constantly not achieving their levels of satisfaction.

The value of client satisfaction measurement

By measuring your client satisfaction levels you will have an important measure of

❑ future client loyalty;

❑ your reputation;

❑ vulnerability to losing future work to your competitors.

To increase the value of these surveys you should

❑ act on the findings with bespoke action for each client;

❑ increase the number of clients surveyed to get a better overall picture;

❑ increase frequency of feedback from clients.

If you have satisfied customers you will find that they

❑ give you more repeat business;

❑ will be prepared to pay more;

❑ they will recommend you to others;

❑ they will cost you less to service;

❑ they will be more profitable.

In fact, they will become your gold clients!

A good starting point to improving service delivery and creating stronger client relationships is to find out how well you are currently doing.

6.9   Third-party survey

One of the best ways to obtain feed back is through a third party so that the client is not aware that it is you that will be receiving the feedback. An example of a questionnaire you can use is shown in **Table 6.4**. It is best to have several parts to the questionnaire, these being the following:

❑ **First part:** Put some research into the survey which may interest the client. You can offer to give him a copy of the results. This gets a better response than just asking the client questions as they all are busy people. You can tell the client that his views, as a key businessperson within the sector, are very important to the survey. Clients love their egos being massaged.

Table 6.4   Client survey

Client survey		
Client	**Feedback from**	**Date**

Part 1: Research

❑ Ask the client his views on a particular topic such as 'What do you believe are the key elements that make a good office design and working environment?'

❑ Make a note of the answers and ask the client to rank them in importance

❑ Have some follow up questions on the same topic and again note the answers and ask the client to rank their importance

❑ Ask the client if he has any other view about the research topic

Part 2: Key players:

❑ Do you know any of the following architects (or engineers etc.)?

❑ Have you worked with any of them?

❑ Are there any others not listed that you have worked with? (add to list below)

	Do you know them?	Have you worked with them?
Firm A		
Firm B		
Firm C etc.		
Your firm		
Others		

Part 3: How did the key players perform?

❑ Of those you have worked with how would you rate them?

❑ If you have worked with more than one of them how would you rank them?

	How would you rate them? Excellent, good, satisfactory or poor?	Rank	Comments against any criteria
Firm A			
Firm B			
Firm C etc			
Your firm			
Others			

Continued

Table 6.4 *Continued*

Client survey		
Client	**Feedback from**	**Date**
Part 4: Survey sponsor		
❑ Tell the client at this point that the survey is sponsored by your firm ❑ Now the client knows it is your firm ask him to rate your firm		
	Excellent, good, satisfactory or poor? (also add any comments)	
How well did they understand your requirements?		
How well did they perform?		
Other questions and comments		

❑ **Second part:** Ask the client if he is aware of any of the firms from a list of professional firms. The list contains your firm plus your rivals within the industry. Also ask the client to add to the list any not mentioned. Make a note of the firms known and those added.

❑ **Third part:** Now ask the client to rate the firms on the list from, say, 1 to 4. Always use an even number or else the client will be tempted to use the middle one when not so sure. Have several criteria to evaluate. Also, ask the client if there were any criteria that you may have missed out and would like to add.

❑ **Fourth part:** This is where the interviewer advises the client that the survey was sponsored by your firm and would he be willing to answer some more specific questions particularly related to your firm? If you want to know specific feedback from this client then ask the client if is he happy for the information to be passed on as being from him. If this is a wider survey with many clients being surveyed, then you can tell the client that his views will be relayed to you, but all feedback will be confidential from the point of view of who said what.

The benefit of a third-party survey, where the sponsor is unknown, is the client will be more open and the person asking the questions will be less biased and will record all feedback given (with no editing!!).

Depending on the length of survey (should not be more than 15 minutes), the location of clients to be interviewed and the budget, the survey can be by telephone or one-to-one interviews. If the budget allows, then face-to-face appointments are better, as the clients tend to give more information and feedback.

If the survey is being carried out by telephone then an appointment is made so that the client is ready for the survey and has allocated sufficient time.

6.10   Direct questionnaire

During the project, at key milestone events it is a good idea to send out short feedback questionnaires (see **Table 6.5**). Do not wait until the end of the project, unless it is a very short assignment, to get feedback. If there are any negative comments, you want to have an opportunity to address the situation before the job is finished and, hopefully, turn the client around.

6.11   Client review meeting

There is always a need to telephone the client or meet the client with the specific task of obtaining feedback. Clients like to be asked. They want you to care about the level of service you are offering them.

Holding a client review meeting is one of the most important activities you can do with a client. These meetings should not just be confined to how the current project is going but can, and should be, expanded to incorporate the following:

❏ What are you doing well and how can you improve it further?

❏ What are you not doing so well and how can you improve it?

❏ Advise the client if any of his procedures are causing problems and discuss if these could be modified for the project benefit.

❏ Investigate forthcoming work within the current project to antici-pate any problems or bottlenecks.

Table 6.5 Client feedback questionnaire

<table>
<tr><td colspan="5" align="center">**Client Feedback**</td></tr>
<tr><td>**Client**</td><td colspan="4">**Project**</td></tr>
<tr><td>Feedback from:</td><td colspan="4">Job no:</td></tr>
<tr><td>Feedback noted by:</td><td colspan="4">Rating: cross/tick in relevant box</td></tr>
<tr><td>Date:</td><td>Excellent</td><td>Good</td><td>Satisfactory</td><td>Poor</td></tr>
<tr><td>1 How well did we understand:
❑ The nature of the work?
❑ The project objectives?
❑ The terms of reference?</td><td></td><td></td><td></td><td></td></tr>
<tr><td>2 How well did we perform:
❑ Technically?
❑ In liaising with you?
❑ In reporting results?</td><td></td><td></td><td></td><td></td></tr>
<tr><td>3 Did we deliver results:
❑ On time?
❑ To cost?
❑ In full?</td><td></td><td></td><td></td><td></td></tr>
<tr><td colspan="5">4 Did our contribution add value to the project or fall short in any way?</td></tr>
<tr><td colspan="5">5 Did we demonstrate an adequate level of commitment to your project?</td></tr>
<tr><td colspan="5">6 Other comments:</td></tr>
<tr><td colspan="5">Management comment:</td></tr>
<tr><td>Action required:</td><td colspan="2">Who by:</td><td colspan="2">By when:</td></tr>
</table>

❑ Bring to the client's attention any benefits that you are bringing to the client's organisation, which they may not be aware of.

❑ Identify any other opportunities that may be coming out from the client.

❑ Advise the client of other projects, which have had distinct client benefits, which may, in part, be transferable to the client organisation. Pursue any possible ideas arising form this discussion.

❑ Enquire if there could be any referrals the client could make to other parts of the client organisation or to other businesses.

❑ Capture and address any other issues that the client may have.

At the end of the review meeting, just summarise any action you have arising from the discussion. Confirm any deadlines attached to those actions.

After obtaining feedback

After obtaining feedback it is important that you

❑ praise those who are performing well;

❑ talk to those who are not performing well to see what the problems are and look to resolve the issues;

❑ put in place a recovery plan/action plan to bring back the project on course if feedback highlights problems

❑ Arrange to see (or contact) the client to advise him that you have listened to his feedback and of what action you have put in place.

6.12   Lessons-learnt workshop

At the end of the project, it is worthwhile to arrange a lessons-learnt workshop meeting with the relevant parties to discuss the following:

❑ How was the project implemented?

❑ What were the good points?

❑ What were not so good?

❑ What would you do differently and why?

A lessons-learnt workshop gives valuable feedback to all those concerned so that they can do better next time with the same client.

Also any improvement will be good for the benefits of all clients, a form of continuous improvement.

The other vital benefits from a selling of professional services point of view are as follows:

❑ The client sees that you care and are willing to learn from his assignment.

❑ It keeps you in contact with the client, especially if you hold the workshop, say, 3–6 months after the project is complete. That time allows the client to settle into the new building or see the benefits of the assignment, if not a building.

❑ Helps strengthen ties between various people within the client organisation and your organisation.

If the workshop has some meaningful outcomes, then you can write it up as a case study and share it with your client. The more interaction between you and the client the better, especially after the project is complete, so you are around to pick up any follow-up work. There is a need to be regularly in touch.

The need to develop multi-level contacts throughout the duration of the project is discussed in Section 6.14 but there is also a need to continue those contacts after the project has been completed.

6.13   A client expectation charter

From the client feedback and workshops on how the relationship could be improved, it would be good to draw up a client promise or expectation charter. Different clients value different things. Some of the services that clients might want to pick up in the client charter may include the following:

❑ A 24-hour access to project status. This could be with an extranet that the client can exclusively enter and check the status of current work, be it drawings or reports.

❑ All the initial promises that were 'sold' at the outset of the commission being delivered.

❑ More face-to-face meetings.

❑ Less face-to-face meetings but more conference calls.

❑ Telephone calls returned within a few hours or within a day.

❑ Action points being cleared by the agreed dates.

Never assume what is important to the client. His needs may change as you pass through the life cycle of the project.

Once you have established the needs, write them down and share them with all the team, including administrative staff, and the client. Draw up a 'Client Expectation Charter' and review it and monitor how people are performing on a regular basis. These could be your key performance indicators (KPI), which is sometimes established by the client at the outset of an assignment.

Service delivery review meetings

Having agreed to review your service delivery with the client, make a feature of it. Clients think highly of service providers who monitor their service delivery. It shows that you are listening to their feedback and you need to make sure that the clients know that you have listened and acted accordingly.

Hold a review meeting with the client. The format could be as follows:

Review of current services

❑ What is good?

❑ What is not so good?

❑ What action is required to improve?

What changes are there within the client organisation that you need to be aware of?

❑ Are there different procedures?

❑ Are there different priorities?

❑ Are there opportunities to assist on other assignments?

What can be gained from the service delivery so far?

❑ Can we write up a case study for client circulation and ours?

❑ Can we promote externally what we have achieved together? (Clients also want to be seen in a good light among their peer group.)

Arrange for the next meeting and interim progress reports as agreed.

6.14   Building multi-level contacts

During the early relationship-building process, it is important to strike up relationships with the various client contacts who can influence your appointment and give you repeat work.

Initially, there will be one or two people within the client organisation that you will be building good rapport with. You and your client contact will have other people who, as yet, have probably not met. This can be best shown in **Fig 6.4**, which looks like an hourglass with the main interface just limited to two people, one from each organization.

You and your client contact are at the narrow interface between the two organisations and the ongoing relationship is maintained from that limited interface. The downside to this is as follows:

❑ If you or your client contact move on then the relationship between the two organisations is greatly weakened and put at risk.

❑ You or your client contact might be the right people to start the relationship and get a project on the road, but might not be the right people to organise the implementation.

❑ You or your client contact might not have sufficient time or skills to remain the appropriate interface for the successful completion of the project.

Therefore it is important to start building several relationships between the two organisations. This can be shown by converting the

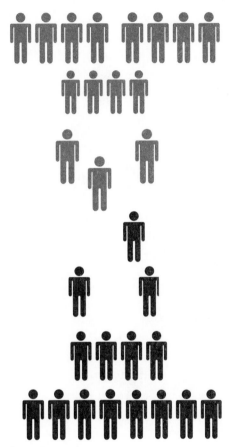

Figure 6.4   Limited team interaction.

hourglass shape into a ball where everyone has easy access to their opposite number and others in the team as shown in **Fig 6.5**.

Apart from helping to resolve and limit the potential problems outlined above, the closer multi-level relationships will also do the following:

❏ Let the two organisations understand each other better and therefore help communication and delivery of the project.

❏ Each team member can contact their opposite number and build their own departmental interrelationships.

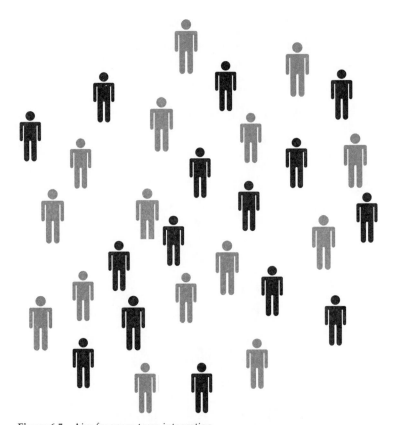

Figure 6.5 Aim for more team integration.

❑ Any potential conflicts can be identified at an early stage and resolved quickly before they develop into problems.

There are various ways that multi-level contacts can be developed and these are outlined below.

Introduce someone else at every opportunity

When you visit the client, take the opportunity to take with you someone else from your organisation. If the selection of the person is relevant to the meeting then that's better. Your colleague can be introduced as someone who will be looking to work on the project and is accompanying you to get to know the project and client. Your client might then take it upon himself to show your colleague round or introduce him to other members of the client team.

When taking up this initiative, let the client know beforehand and perhaps even suggest that the colleague meets his opposite number within the client team.

Show them around your office

In this busy world more and more meetings are being held in the client's offices. Try to tempt the client to visit you in your office and take the opportunity to show him round at the end of the meeting.

Beforehand you will need to brief people whom you will 'bump' into along the way. If possible, stop by their workstations and introduce the client. Ask your colleagues to briefly outline what they are working on (subject of course to client confidentiality). This way you are marketing to your client in a soft sell. It's just 'incidental' conversation, but you have orchestrated it to show off your people and your skills and experience.

Ask to be introduced to your client's colleagues

When visiting your client ask about his colleagues. Ask to be introduced to key people who will have a part to play in the project. Meet the finance man. This will be very useful if later you are chasing a late payment. Speak to some of the end users. Get them on your side. Listen to their comments. If a particular comment comes up in conversation, volunteer a visit by one of your team to help clarify a point or take a brief about a particular aspect.

Hold pre- and post-project social gatherings

Take the opportunity to celebrate the relationship between the two organisations. At the beginning of the project hold a pre-meeting workshop when the two teams meet up and tell each other what they do. They can also outline any worries they may have or things that have gone wrong in the past and could be revised for this project. End the workshop with a social event. Depending on the budget this could be a buffet, dinner, quiz, challenge, etc. If you do have a competitive event, then don't have the two organisations head to head but mix up the people, so teams have representatives from both sides.

Hold value engineering workshops

This not only has a financial benefit to the project but also gets both sides working to a common goal of looking carefully at various aspects of the project. It really is a good idea to bring in a trained facilitator for this. This exercise will also send signals to the client that you are happy for the project to be modified so that it satisfies the client's needs at the appropriate price and quality.

Market the project with the client

Some projects will be newsworthy. At every opportunity, take time to think ahead and plan a publicity campaign (see **Table 6.6** for possible milestones). Obtain the client's approval. Quite often, the client will be more than willing to show off his project. Get to know the client's marketing department or head of PR; they can be very useful in providing background information and facts and figures and even pictures.

Becoming more integrated

With time, you should be building an increasingly stronger relationship with your client. The relationship may go through the following stages:

Supplier: Often the first assignment has been won in competition and may have been based on a mixture of price, experience and capability.

Preferred supplier: Once you have worked with the client and have delivered a good service and met all the client expectations, then the client becomes more relaxed and confident in using your services again. You then move up the relationship ladder to become the preferred supplier.

Partner: With time and with experience of the client and the client's working methods, you may be elevated to partner status. This is

Table 6.6 Example of raising profile opportunities throughout the course of a project

Raising profile opportunities

Office	Job title	Client	Dates	Customer feedback						Project events				Press release	Photographs	Job sheet	Competition entry	Golden letter	Score
				Press release	CRM					Site board	start	Top out	Hand over						
				1	2	3				4	5			6	7	8	9	10	
	Warehouse	Brown & Co.	Start: 08/08 Finish: 08/09	X	X	X	X		X	X	X	X							8
				A 10/08/08	A 04/08/08	A 08/08/08	A 01/01/09	P 01/05/09	A 10/08/09	A 20/08/08	A 08/08/08	A 16/03/09	P 20/08/09	P 30/08/09	P 20/08/09	P 10/09/09	P 06/02/10	P 10/11/09	

Action dates
A:Actual
P:Planned

219

when the client will see you as providing true added value and will become reliant on you. The client might also call you in earlier in the process to assist him with his building or site strategy.

Large clients with an ongoing portfolio of work will not want to go out to the market each time as this would be too costly and have a risk of entering into a fresh relationship and the need to go through a learning curve. Also, if the client is the local or national government, they are obliged to advertise within the OJEU (Official Journal of the European Union). We have seen the formation of framework agreements to overcome this.

A framework agreement will go through the usual procurement process of being advertised, expressions of interest, pre-qualification and invitation to tender. The tender will be based on a large workload being made available to a limited list of selected contractors or service providers. The framework might

❑ offer each framework contractor, or service provider, work in rotation;

❑ offer selected contractors, or service providers, projects from within pre-determined values, which match their resource and experience and price levels.

The client sees benefits that he has a degree of competition and is able to share work among a select number of providers. If enough work is given out at regular intervals, the providers are able to maintain teams which can move from one completed project for the client to the next new one.

These frameworks will often last for 3 or more years.

Winning additional work from clients

Once a project has been secured there is an opportunity to do the following:

❑ Sell more of the same skills to other parts of the client organisation. You might be dealing with one part of a client, perhaps their research division. Try to get introduced to other parts of the business, perhaps manufacturing. Also if you are working for one region, try to get work in other regions. Use your existing client contacts to get you into these other areas.

❑ Sell different skills to existing contacts. You might be providing a specialist service and might be able to sell different services to the same client team.

Use your client contact time profitably to find out about other potential work within the client organisation and try to get introduced to more people. This will be so much harder after the project is completed because opportunities to meet the client will be less. At least during the current project you can be making sure that your client is happy with current service levels, and if he is, it is so much easier to ask for those introductions. Within large client organisations there may be opportunities to sell more of your services as identified in **Fig. 6.6**.

Succession planning

When developing long-term relationships with a client, it is important to recognise that the client belongs to the organisation and not the individual. Too many businesses have seen long-established relationships come to a rapid end when a key individual from either the client side, or within the professional firm, leaves the business. This might be due to retirement, promotion within the client team or moving onto new ventures.

It is therefore vital that as soon as the client relationship is established or when it is being enhanced then

❑ more people are brought in to meet the key client team and build strong links to establish succession;

❑ more contacts are made with the client's side, especially with potential successors within the client team;

❑ develop a knowledge base where all important information about the client is recorded so no key information is lost.

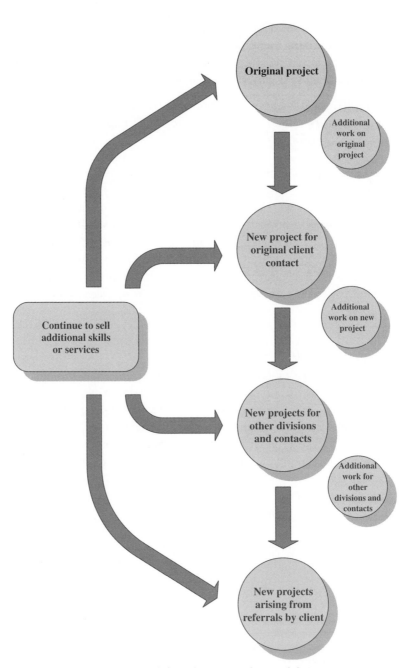

Figure 6.6   Targeting more work from the existing client and their contacts.

Once there is a hint that the client is about to move on to new pastures, bring into place a plan to make sure that your team is well embedded within the client organisation. There will be risks if a successor comes in from outside as he may be tempted to bring in his professional team that he has used before. Equally, you want to also follow the outgoing client contact to his new venture and create new opportunities there.

If you are made aware of change of this kind then you need to do the following:

❏ Find out who has been appointed to replace your client contact.

❏ Do some research on the incoming new boy; find out who he has been using and how vulnerable you are.

❏ Often there is an overlap between outgoing and incoming people, so get your old contact to introduce you to the new person and request him (if you know him well) to sing your praises.

❏ Also, agree to visit the outgoing client in his new post as soon as he has got his feet under the table. Agree with him on what a reasonable time would be and make that visit.

If you follow these steps, then not only do you protect your relationship with the existing client organisation but you also enlarge your potential client portfolio by keeping in touch with the outgoing client.

6.15 Client loyalty

Clients become and stay loyal because they receive, and are aware of receiving, the best value service. Therefore, as mentioned previously, the professional service provider needs to continuously monitor and revise the service delivery he is providing.

Even if the client was satisfied 5 years ago with the delivery you may find that the client has

❏ matured and is in need of a better service;

❏ been experiencing the same, old service and perceives complacency setting in;

❑ your competitors bombarding him with what he perceives is a better offering.

Improving client loyalty can therefore be enhanced by addressing the above points.

Increase client loyalty

To increase your chances of client loyalty you need to do the following: Make them want to use your firm.

❑ Give them that extra service.

❑ Increase client contact at all levels.

❑ Increase the personal as well as business relationships.

Enhance your service offering:

❑ By increasing your knowledge of your client and his market or industry

❑ By getting to know more about the needs of the client organisation and monitoring how these needs change with time

❑ By getting to know the decision makers, influencers and stake-holders

Keep in touch to pursue the next assignment.

❑ Get to know all the job givers and key decision makers.

❑ Find out and track potential projects on the horizon.

❑ Be there when the client switches from 'not buying' to 'buying' mode.

❑ Ask for the assignment, at the right time, without the 'hard sell.'

Setting goals and objectives

Surveys of businesses that have high client retention have shown that they have strategies for setting goals or objectives for their client relationship. Goal setting is a powerful technique that can bring about results. The technique can be used for various stages outlined in this book. You can set goals for the number of calls you make to clients and how often you keep in touch. Also set goals for how much income you want to achieve per sector or within a specific period. Setting down goals will help you concentrate and focus precisely on what you want to achieve.

One of the best pieces of research in this area was at Yale University where graduates were surveyed in the 1950s and again 20 years later. The research showed that 3% of the graduates, who had set themselves goals, were worth more than the other 97%, in terms of wealth, put together.

To increase your chances of making your goals succeed try to make them SMART.

- ❑ **S** is for specific. A well-defined goal (e.g., you want to obtain 40% repeat business from existing clients).

- ❑ **M** is for measurable. If you can measure the progress, you can make changes to keep you on track. It is also good to put down key milestones along the journey to help monitor progress.

- ❑ **A** is for achievable.

- ❑ **R** is for realistic.

- ❑ **T** is for timed. Your targets should have timings so that you can gauge how you are progressing (e.g., you want to increase from the current 10 to 20% within 1 year and then 30 and 40% at the end of the next 2 years).

Do not set yourself too many goals; you must give yourself a chance of achieving them. It is also best to prioritise them. Perhaps target your key gold clients first and then the silver and if you have time try to convert the bronze to silver. Also, it is best to tackle the easier goals first so you have some early success.

Write your goals down and define how you are going to measure your progress. You will also need to know when you have achieved your goal rather than leaving it open ended. You should not have a goal that relies on luck. If you base your goals on personal performance or skills or knowledge to be acquired, then you are able to keep control over the achievement of those goals.

Goals should not be set too low because of the fear of failure or just wanting to take it easy. The goals should also stretch you within reason.

6.16   Obtaining referrals from clients

One of the best endorsements you can have is from a current or past client. In addition, your client will probably belong to his own networks of like-minded people and with people with similar roles within similar sectors.

If you are confident that you are, or have, delivered a good service ask your client if he can give you some names for you to pursue. You could ask your client direct or as a question on a feedback form. Also you could ask them who they collaborate with, such as trade associations; this, in turn, may yield some fruitful contacts for you.

Once you have established a list, and have the client's approval to use his name, contact him offering your services and try to arrange a meeting to explore opportunities. A letter of introduction could be on the follolwing lines:

Dear Mr Bart,

Richard Huggins of Barker & Conlan Health Products Ltd has suggested that I write to you to let you know some of the reasons why he decided to switch to our firm of architects for his new production facility.

Mr Bart has advised me that you are also in the health products business and might welcome a meeting to enable us to show you how we were able to provide Barker & Conlan an efficient, cost-saving processing building. Some of the features we incorporated might indeed be of benefit to your business.

I will telephone you in a couple of days to arrange, if you agree, for an appointment to visit you.

Yours sincerely

Also, if possible, obtain from your client a letter of endorsement that you can copy and attach to your letter to his contacts.

Summary checklist

By the end of Stage 6 you will have a strategy of delivering added value and securing repeat business. Your strategy will include

❑ preparing a project client plan;

❑ putting in place total office participation in selling to existing clients;

❑ managing all the service 'touches' experienced by your clients so that they are pleasant experiences.

To maximise future work from existing clients you will have to put in place

❑ a client account management system;

❑ maximum available time and resources to focus on the best possible opportunities;

❑ client account teams.

And you will have been investigating how to

❑ create new relationships within the client organisations to increase chances of future work;

❑ obtain referrals from your client and follow up on the new contacts to pursue additional work.

In respect of service delivery, you will put in place a strategy to make sure the clients are happy with the services they are receiving and making sure your competitors are kept out. You will have

❑ put in place a client satisfaction survey;

❑ regular client review meetings;

❑ acted on the feedback obtained.

Stage 7: Building credibility

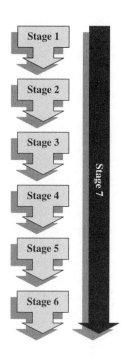

7.1 Credibility through sector knowledge

7.2 Response to requests for information

7.3 CVs

7.4 Keep records of your experience

7.5 Project sheets and case studies

7.6 Using e-mails

7.7 Your website

7.8 Intranet and extranet

7.9 Press releases

7.10 Newsletters

7.11 Research

Stage 7: Building credibility

7.1 Credibility through sector knowledge

Most clients will prefer to select professionals who predominantly work in their sector rather than general practitioners. Clients usually believe that their sector is special and will require a specialist's knowledge and expertise. For this reason, I am not too keen on the general office brochure. If you put all your experience and skills sets together in one document you will come over as a general practitioner. Unless, that is, you only work in one sector.

It is much better to bring together all your sales materials into market sectors. In categorising your work, it is the categorisation of the project rather than your client that is important. However, some projects can fall into two or more sectors. For example, a sports hall for a large school could appear in both the sports and leisure category and in the education category. But the office building for the pharmaceutical conglomerate would sit within the commercial office sector rather than the pharmaceutical sector. You could still put the office building in both sectors if, say, the methods of working, e.g., scientists' writing areas, could be very specific in the pharmaceutical sector and varies from general office space. At the end of the day make your experience work for you and, by all means dissect a project and put the various components into different sectors.

When selling into a sector you will need to fine-tune all aspects of the range of information to show your expertise in that sector.

Information for sector-specific selling

An outline of the various components that could come together to help sell your professional services to a specific sector is as follows. (You may need to obtain client approval for some of the information you wish to incorporate.)

Company introduction

❑ When formed.

❑ How long in the specialist sector.

❑ General statistics on work in the sector (e.g., number of projects, total build value, diverse geographical spread, new build, refurbishment, etc.).

Service offered

❑ The complete range of professional services offered.

❑ The type of work undertaken (master planning, feasibility, full design, site administration, etc.).

❑ Your people. Key experts, CVs which feature only the sector-specific projects, or specialist skills that would be required in that sector.

Your client list

❑ A selection or full list of clients you have worked for in the sector.

Project list

❑ A list of projects you have worked on within the sector. Include project title, client name, location and possibly the project value and when completed.

Project sheets and case studies

❑ A selection of all the project sheets for the sector and some case studies.

The selection should be suited to match the project you are pitching for.

Papers and articles

❑ Reprints of articles and papers delivered to demonstrate your 'expert' credentials.

Awards and endorsements

❑ List any awards and significant achievements within the sector or
relevant to the services that will be provided, along with endorsements and statements from satisfied clients (if from within the
sector then even better).

7.2   Response to requests for information

Sometimes you may receive telephone calls or e-mails requesting
information. Don't just send out the information without making

Table 7.1   **Follow-up to a request for information**

Follow-up to a request for information
Request from:
Name:
Company:
Address:
Tel No:
E-mail:
Information requested:
Questions to be asked:
How did you hear about us?
Do you have a particular project in mind?
Would you like one of our directors to contact you?
Would you like to arrange to meet one of our directors?
Date of information requested:
Information sent:
Copy of this form forwarded to:
Follow-up action:

some enquiries. You need to make sure that the information you will be sending is applicable and you need to find out why the information is being requested. If the request is from a potential client you need to be able to follow up. Your marketing department or the person dealing with these enquiries should find out as much as possible. **Table 7.1** is an example of a follow-up form, which could be used to record the relevant information and help monitor any follow-up action. This form could also be adapted to be used on your website should requests originate there.

In addition, have some overview sheets, or mini brochures prepared that could be sent to students, professionals looking to make a career move, or journalists. Alternatively, have this information downloadable on your internet site.

7.3   CVs

In the professional services world clients are selecting people, who are also your main assets. Therefore you need to maintain a CV bank (see also Section 5.10).

Use a photograph (professionally taken) of the person and keep on file as much information as possible about the person. Include the previous firm's experience as well as experience within your firm.

Have someone in the firm be in charge of updating CVs. Hand out current CVs to the employees every year so that they can update, and also give them a chance to adapt the CV for the specific pitch being pursued.

There may well be various forms of CV on file. For each individual create at least two CV formats. This will include:

❏ **Full detail.** The full CV (rarely used externally but used for extracting information for other specific CVs.)

❏ **A sector CV.** Tweaked for sector-specific detail. This can be a full sector-specific CV or an abridged version, depending on the documentation being put together.

7.4   Keep records of your experience

Potential clients will want to be reassured that they are appointing someone who can deliver the project they have in mind. The main way clients can establish your capability and credibility is to examine your past experience. Evidence of past work can be demonstrated by

❑ marketing material such as project sheets, case studies and published articles about your work;

❑ favourable third-party endorsements such as client references and favourable reviews in the trade press;

❑ site visits to see the project for real.

It is amazing how little information is kept by professionals about the work they have done. They always seem to be too busy to keep appropriate records. There is a need to keep a record of key information about a project. At least if the information is kept, then appropriate marketing material can be created incorporating facts at a later date. It is so difficult to write an interesting copy without a good sprinkling of facts.

Not all businesses can afford full-time marketing people to keep on top of the fact-gathering exercise. Even if a marketing professional is employed they will often have to rely on the team members to provide the information.

The easiest way to resolve this problem is to make it part of the procedure that the job cannot be closed down or archived until all the key information is gathered. Key staff are likely to move on and with them goes the knowledge about the project. So create a system, or culture, for harvesting project information during the course of the project lifetime.

Key information should include the following:

❑ Client details, key contact details.

❑ The team. Who in the firm worked on the project, other team members from other firms, roles and responsibilities?

❑ Purpose of the project.

❑ What was required?

❑ How did you achieve the brief?

❑ Any added value, benefits (define and quantify).

❑ Key facts, figures and dates.

❑ Photographs of the completed project and any good visuals and graphics that may have been produced. Possible photographs during construction especially if showing off a new or unusual methodology.

❑ Client endorsements ('golden letters').

It is always better to record too much information and be able to edit it rather than having to struggle to find information. This becomes especially difficult if key people have left the firm and records have been destroyed.

7.5 Project sheets and case studies

It is useful to create information in varying lengths, so that you have a ready-made copy to suit all eventual needs. These would include the following:

❑ A paragraph which summarises the project

❑ Several paragraphs that can be pasted into a document to give an outline of the project

❑ A project sheet which would be of A4 size (landscape or portrait) and incorporating illustrations

❑ A case study which could well be of several pages and gives a more detailed explanation about the project

With the above information, you will be able to demonstrate to the client that you are able to undertake his project. But you must be prepared to be very selective and even rewrite project sheets to suit the project you hope to secure.

If there is something which is very important in the proposed project then feature that aspect in the project sheet. Don't dilute the impact by writing about aspects of the project which are secondary or of little relevance. Remember that the client may have many other teams pitching for the project, so you need to grab his attention with facts, features and benefits.

This does take time and effort and, all too often, the busy professional will cobble together information in the hope that some of it will be relevant.

If you have been given a brief, or have made notes during a briefing session, then you must do the following:

❑ Feature the information in the same priority as the client does. If the client has spent most of the briefing session saying he wants a low maintenance, energy-efficient building, then don't start off with big chunks of text talking about how fast the building was built and how unique the structural solution was.

❑ Use the client's terminology wherever you can.

❑ Keep your text short and to the point. Make it easy to read and use short sentences or even bullet points. You need to maintain their interest.

❑ If appropriate and of interest, convert facts to more manageable information. Rather than giving the acreage, you might say it was equivalent to so many football pitches. Or if the height is a key feature, why not show the project against a well-known landmark such as Nelson's column or the Eiffel Tower.

For example, you could say:

The warehouse was a large single space which was the size of ten football pitches.

The clear height within the warehouse was more than three double decker buses.

If you take the time and effort to make the project sheets bespoke then you will stand out from your rivals. I have seen so many professionals just 'cherry pick' from their marketing material and then expect to get shortlisted without making the submission bespoke. It would be much better to only pursue projects that you think you stand a good chance of winning and spending the time on those. Why not just chase one in five of the ones you are chasing now and increase your success rate by producing better submissions?

Use your project sheets as door openers

If you get into a routine of creating project sheets why not send them to clients and potential clients just to let them know what you are up to. This helps to

❑ keep your name in front of the client;

❑ show that you have a varied client list;

❑ show that you are at the forefront of design, technology or whatever that may be relevant and of interest.

7.6   Using e-mails

If you have the client's permission to e-mail him information, then why not drop him an e-mail with a short paragraph about the project and give a web link so, if he is interested, he can click and read more when he has time. Be careful about sending images. These can

❑ be blocked by some organisations;

❑ be too large and annoy the client by filling up his e-mail memory allocation imposed by his IT department;

❑ take too long to load up that the client gets fed up and into the wrong frame of mind.

Alternatively, you could use your e-mails as a news update system. Just imagine how many e-mails leave your office every day; if you could harness each one to deliver a message, you will be raising your profile with minimal input.

Consider placing a single eye-catching sentence below the main body of the e-mail but above the legal disclaimer section. The following are some examples of messages.

To demonstrate how **successful** you are

> *Wren & Barry have just been selected to design the new headquarters for Steel and Carbon Industries.*

To demonstrate your **good service** delivery

> *New headquarters for Steel & Carbon Industries completed ahead of time and within budget.*

To demonstrate how **talented** you are

> *Design award presented to Wren & Barry for their stunning new headquarters building for Steel & Carbon Industries.*

To demonstrate your **green credentials**

> *New headquarters building uses sun and wind to reduce running costs by 30%.*

To demonstrate your **knowledge** base

> *Wren & Barry host seminar on latest thinking on improving the office environment.*

To demonstrate your **growth** and succession planning

> *New associate director joins Wren & Barry to develop their interior design department focusing on the integration of latest IT technologies.*

To demonstrate your position at the **forefront of the profession**

> *Latest survey and article by Wren & Barry shows clients want more flexibility in office designs.*

To demonstrate **up-to-date thinking**

> *Wren & Barry agree with latest government thinking on insulation standards. Click here to read how this may impact on your next project.*

The above examples show that you can raise your profile by homing in on the various stages and key aspects of a single project. If you do this with several projects you can really raise your profile and appear to be very busy and successful. Clients are attracted to busy successful people.

You need to be careful to not just rely on one or two projects. Most projects will have some newsworthy information you can use to create a single line at the foot of your e-mail. Try to change the message frequently, daily if possible, but don't leave the same message on the system for longer than, say, a week.

7.7 Your website

No matter how small a business you are, you need to have a website. Websites are used by the following people:

❑ Potential clients who may be using the website to help prepare a short list of professional service providers.

❑ Other professional service providers who may be looking for consultants to team up with on projects.

❑ Contractors who may be looking to appoint consultants for a design-and-build project.

❑ Future employees who may be looking at your website to see if your organisation would be a good place to develop their careers. A good website could attract talented employees.

❑ Researchers who may wish to contact you for an article or news story they are preparing.

❑ Competitors who will look at your website especially if they know they are competing with you on a particular project. Likewise you will be visiting their website.

Unfortunately, too many websites are bland and do not help potential clients in their selection process. You need your website to demonstrate the following:

❑ **Who you are.** Provide an overview of the firm.

❏ **Your people.** Show off your employees. Remember people, pick people. You will need to demonstrate a cross section of people; don't just focus on the partners and directors.

❏ **Your experience.** Show off your sector skills with images, key facts and figures and perhaps a few case studies and project sheet adapted to suit the style of the website.

❏ **Your services.** Show off your professional qualifications and full range of the services you offer.

❏ **Your success.** Have a news section for latest news and an archive of past press releases.

❏ **Your knowledge.** Give access to your published articles, papers, etc. Also, within the sector, comment on current issues and thinking and how you have been able to address these. You will have to learn to give some information away to demonstrate your knowledge and to put you ahead of the competition.

Always spend some effort to update your website and tell potential clients to visit your website for information (see e-mails above). Try to avoid the following:

❏ No site or under construction. What message does that convey to your target clients?

❏ Old information. Don't feature out-of-date news stories.

❏ Bland presentation. Make it interesting and easy to navigate and well presented graphically.

Most professional service providers are happy for their website to be found just by feeding in the business name into a search engine. If, however, you want to raise your profile within certain sectors or for particular niche services then consider obtaining good quality advice from web designers.

Your web designer will examine the content and the relevant links to a specific search a potential client may have. Your designer will also make the site 'friendly' to search engines.

In addition, you need to create external links that may divert potential clients to your site. You might be able to get other organisations to feature articles you have written, or for them to review talks that you have delivered. In this situation try to have your website link included, this may entice people to click onto your site. For example, you might be speaking at a conference on the latest thoughts on laboratory designs. The conference organisers may have a website to promote their conference and to take bookings. Your name and, hopefully, website link, could be included in the conference agenda. Therefore delegates considering booking may visit your website. You will be surprised how long these conference links remain on the internet after the conference is over.

7.8   Intranet and extranet

The intranet is used internally and cannot be accessed externally. Use it to keep your employees informed, as they are your ambassadors. Tell them of the latest recruits, job wins and social events. Your employees will rarely visit your internet site, so feed them information direct to tempt them to visit the intranet. Consider using software which tells staff about new information on the intranet when they switch on the PC in the morning.

The extranet is a secure part of your website where access is given to selected people such as clients. There might be parts of the extranet where only one client can access so he can check progress on his project.

Use extranets to depart information and intelligence to your clients or target clients, using a password system. It can also tell you what your clients find interesting.

7.9   Press releases

The purpose of the press release is to raise the profile of your organisation. Therefore you need to target the press release to achieve different levels of awareness. Consider the following:

❑ **Local press.** This is especially important if your workload comes mainly from the local community. The important feature to focus on is the human-interest aspect, especially for the general pages. If your local or regional paper has a business section, consider highlighting how your project has benefited your client's business.

❑ **Sector or trade press.** This is always a good area to target. It is more likely that your future clients will be reading these types of magazines or journals.

❑ **Peer press.** This might be good for the ego but probably few future clients will be reading your trade or professional journals. However, there is the benefit that future employees might be tempted to apply to work for you, especially if your name keeps coming up in the professional journals.

When issuing a press release, make sure of the following:

❑ It is on special headed paper with 'Press Release' in bold print at the top so that it can be directed to the news desk.

❑ It is current news.

❑ It has interesting information relevant to the publication and its readership.

❑ Make it interesting, and outline the story in the first paragraph.

❑ It is easy to read and understand. Don't use jargon.

❑ Don't waffle, just report the facts.

❑ Use one A4 sheet, maximum of two sheets, with your contact details plus a few paragraphs describing your firm by way of background at the end and not as part of the story.

You may wish to consider a PR agency if budget permits. The benefits are that they will know how to write good copy, can present it in a journalistic way and highlight the issues that will be of interest to the readership. Also, the PR agency may well know the news editor and be able to lobby them to give you a better chance of being featured.

Consider the various press release angles for a story announcing the completion locally of a new laboratory within the pharmaceutical industry.

To the local press (general)

New laboratory creates jobs for science graduates.

To the local press (business pages)

New laboratory secures future for local business within the pharmaceutical sector.

To the pharmaceutical press

New flexible 'hot labing' gives researchers access to non-dedicated laboratory space on a 24/7 basis, reducing capital expenditure and maximising usage.

To the design press

Wren & Barry design state-of-the-art laboratories with built-in sustainable features.

7.10 Newsletters

If you have enough news to produce a newsletter then this is a good way to keep in touch with your clients, past clients and potential clients. There are a few points to consider:

❑ Have enough news to provide a newsletter. You don't want to feature only one project.

❑ Don't have a bumper first issue; subsequent ones get smaller and smaller and fizzle out. Perhaps consider an annual review rather than an ambitious monthly or quarterly production.

❑ Clients do not have much time to read newsletters, so you need to differentiate yourself somehow. It must be professional, attractive and well laid out. Give information away, which the client might value.

❑ Will you be sending the newsletter out as hard copy or electronically? If electronically, some client's servers might block it due to file size or being considered as spam.

❑ Put your newsletters on your internet site as this again will build up your credibility.

7.11   Research

Bigger professional firms often use research to raise their profile, but there is no reason why smaller firms can't do so as well. If you do the third-party client survey, as outlined in Section 6.9, then there is scope to use your findings to publish an article.

Research needs to be

❑ topical;

❑ thought provoking; and

❑ useful.

Again put all your research on the internet.

If you are able to, offer to send a full report to anybody who wants to have it. Give your details at the end of the article. Needless to say, if people request a full copy then after a few days, follow up and see if there are any project opportunities.

Summary checklist

By the end of Stage 7 you will have in place procedures to be able to build on your experience and develop

❑ records of your experience on a job-by-job basis;

❑ record of your experience on a sector-by-sector basis;

❑ project sheets and case studies;

❑ continuously updated CVs.

You will have also considered how you can continuously promote your services by using

❑ your website;

❑ e-mails that can also direct clients to your website;

❑ press release;

❑ articles;

❑ newsletters;

❑ research to publish.

Further reading

Ashton, R., 2004. *How to Sell*. London: Hamlyn.

Fisher, F., Ury, W., Patton, B., 1991. *Getting to Yes*. 2nd ed. London: Random Century Limited.

Hazeldine, S., 2006. *Bare Knuckle Negotiating*. Great Britain: Lean Marketing Press.

Hazeldine, S., 2006. *Bare Knuckle Selling*. Great Britain: Lean Marketing Press.

Heppell, M., 2006. *Five Star Service One Star Budget*. Great Britain: Pearce Education Limited.

King, G., 2007. *The Secrets of Selling*. Great Britain: Pearson Education Limited.

McCarthy, P., Hatcher, C., 1996. *Speaking Persuasively*. Australia: Allen & Unwin Pty Ltd.

Payne, P., Christopher, M., Clarke, M., Peck, H., 1998. *Relationship Marketing for Competitive Advantage*. Oxford: Butterworth-Heinemann.

Johnson, L.J., 1994. *Selling with NLP*. London: Nicholas Brealey Publishing Limited.

Thompson, P., 1996. *The Secrets of Communication*. Great Britain: Simon & Schuster Ltd.

Vicar, R., 1993. *First Division Selling*. London: Kogan Page Limited.

Walker, K., Denvir, P., Ferguson, C., 1998. *Creating New Clients*. London: Continuum.

Walker, K., Denvir, P., Ferguson, C., 2000. *Managing Key Clients*. London: Continuum.

White, L.W., 2004. Sales game: *A Guide to Selling Professional Services*. Bloomington: AuthorHouse.

Wilson, C., 1998. *Profitable Customers*. 2nd ed. London: Kogan Page Limited.

Index